t Version

GENERAL ZOOLOGY

LABORATORY GUIDE

Eleventh Edition

Charles F. Lytle
North Carolina State University

J.E. Wodsedalek
Late of the University of Minnesota

 Wm. C. Brown Publishers

Book Team

Editor *Kevin Kane*
Developmental Editor *Carol Mills*
Production Editor *Kay J. Brimeyer*
Designer *Elise A. Burckhardt*
Art Editor *Donna Slade*
Photo Editor *Carrie Burger*
Visuals Processor *Ken Ley*

 Wm. C. Brown Publishers

President *G. Franklin Lewis*
Vice President, Publisher *George Wm. Bergquist*
Vice President, Publisher *Thomas E. Doran*
Vice President, Operations and Production *Beverly Kolz*
National Sales Manager *Virginia S. Moffat*
Advertising Manager *Ann M. Knepper*
Marketing Manager *Craig S. Marty*
Editor in Chief *Edward G. Jaffe*
Production Editorial Manager *Colleen A. Yonda*
Production Editorial Manager *Julie A. Kennedy*
Publishing Services Manager *Karen J. Slaght*
Manager of Visuals and Design *Faye M. Schilling*

Cover photo © Ed Robinson/Tom Stack & Associates

ISBN 0-697-05210-9

Some of the laboratory experiments included in this text may be
hazardous if materials are handled improperly or if procedures are
conducted incorrectly. Safety precautions are necessary when you
are working with chemicals, glass test tubes, hot water baths,
sharp instruments, and the like, or for any procedures that
generally require caution. Your school may have set regulations
regarding safety procedures that your instructor will explain to you.
Should you have any problems with materials or procedures,
please ask your instructor for help.

Printed in the United States of America by Wm. C. Brown Publishers,
2460 Kerper Boulevard, Dubuque, IA 52001

10 9 8 7 6 5 4 3

Dedication

I wish to dedicate this eleventh edition to my graduate school mentor and my good friend Sears Crowell, Professor Emeritus of Zoology at Indiana University, for his wit, his wisdom, and his patience.

Contents

8

Platyhelminthes   113

9

Hints for Dissection   131

10

Pseudocoelomate Animals   135

11

Mollusca   147

12

Annelida   161

13

Arthropoda   175

14

Echinodermata   205

Contents

Preface

Each year, new information and new interpretations are added to our knowledge of zoology. New applications of this knowledge continue to be found in human and animal medicine, in agriculture, and in environmental science.

A solid foundation of basic zoology is fundamental for the preparation of students training for careers in biology, zoology, genetics, physiology, medicine, veterinary medicine, agriculture, environmental science, and many other fields. A meaningful laboratory experience is a vital part of this training and furthers the understanding of the structure and function of animals. For more than fifty years, this laboratory guide has assisted students and teachers in their laboratory work with accurate information, clear directions, and quality illustrations. This eleventh edition continues the tradition of excellence by providing a comprehensive guide to the common types of animals representing the major animal phyla. We have attempted to provide a helpful introduction to the major groups and representative forms, incorporating new data and interpretations while retaining important historical perspectives.

Since this book is intended to aid students and teachers in many diverse institutions with different schedules, resources, and preferences, we have chosen to be inclusive rather than exclusive in our coverage. Thus, we have included more material than most zoology instructors will be able to incorporate into their courses. We hope that you will select those chapters, sections, and animal types most appropriate for your course.

Important Features of the Short Version

This short version of the laboratory guide has been prepared for use in a one-semester or one-quarter course in general zoology. It is identical to the complete version except for the omission of chapters on shark anatomy,

perch anatomy, and fetal pig anatomy. The complete version would be more suitable for courses that include laboratory study of one or more of these forms.

Changes in This Edition

This eleventh edition represents a major revision of this zoology guide designed to provide students with the best and most current materials to aid their laboratory studies. We have added 101 new illustrations, many of them in color for the first time. We have attempted to use color illustrations for photographs and drawings where the addition of color aids significantly in the clarity or meaning of an illustration. Some topics and some specimens are best illustrated in color; with other topics or specimens, the use of color adds little except cost.

The 101 new illustrations include 77 new photographs and 24 new or revised line drawings. Many of the new photographs are of expert dissections of representative species done by Dr. Larry Grimes and photographed by Mr. Ken Taylor. Several of these photographs show anatomical details rarely found by zoology students (and many teachers!) and should enhance student laboratory experience.

We have deleted the chapter on animal behavior and shortened the material on chick embryology in chapter 4 to provide space for a new chapter on rat anatomy requested by several teachers. The study of the rat with this new chapter will provide students with an easy and relatively inexpensive introduction to mammalian anatomy and physiology not available in previous editions.

We have followed the suggestions of several users of previous editions by adding new illustrations and expanding the discussion of several topics. Among these changes for this edition are the new material on safety procedures in the laboratory, some simple exercises with the compound microscope, new information on the histology of sponges and cnidarians, and a discussion of the life cycle of the liver fluke, *Fasciola*.

We have also added new exercises on the pinworm, on the garden snail, and on the internal anatomy of the squid and of *Peripatus*. The chapter on the frog has been extensively revised and reorganized, and new material on frog muscles has been added. We believe that these changes and additions will make this book a more valuable aid for students and teachers in their laboratory studies of the fascinating world of animal life.

Basic Features of This Manual

In this eleventh edition, we have retained the basic organization, approach, and educational features of the previous edition. These features are intended to facilitate the work of students and teachers in completing the various exercises.

A Correlation Chart at the beginning of the manual will help teachers and students coordinate the laboratory exercises with several of the major zoology texts in current use. Some important features in the manual include boldface headings within each chapter indicating the major divisions of the exercises, and boldface terms throughout the text emphasizing terminology students should learn and understand. Several chapters provide space for students' drawings of specified structures of animals. The manual also includes several blank tables and graphs for students to complete with data from their laboratory observations and/or experiments.

Every chapter begins with a list of Objectives that points out specific things students should accomplish in each exercise. I have used such objectives in my own teaching for many years and have found that the objectives help focus attention on the important tasks in each laboratory exercise. I also suggest that instructors use the list of objectives when composing lab quizzes to ensure that students understand what they will be tested on and that the tests focus on important concepts, processes, and facts.

Most chapters contain a brief Introduction, with background material on the topic and a Materials List for the exercises. Many chapters have one or more lists of Demonstrations, which are suggested to supplement the main activities of each exercise. Toward the end of each chapter is a list of Key Terms introduced in that chapter; and, with the exception of chapter 9, all the chapters have blank pages at the end for additional Notes and Sketches. At the conclusion of each laboratory exercise, each student will have a complete, consolidated record of his or her lab work; thus, all of the necessary study materials for laboratory quizzes and for review will be in one place rather than in a series of scattered papers and drawings.

Acknowledgements

Once again, we have had the benefit of many teachers who have made suggestions for improvement of this laboratory guide. We continue to solicit such suggestions and comments because they allow us to keep in touch with the changing interests and needs of teachers and students elsewhere. We sincerely appreciate your interest and suggestions.

I wish to thank several colleagues who have contributed to the preparation of this edition. I am particularly grateful to Dr. Larry Grimes, who did many of the dissections, and to Ken Taylor for his excellent photographs. I am also most grateful to the Carolina Biological Supply Company and several of their staff members who have helped in many ways in the preparation of this edition. I wish to thank especially Dr. Raymond Flagg, Dr. Ronald Hammond, Mr. Daniel James, Mr. Roger Phillips, and Mr. Wesley Reynolds.

Other colleagues who have provided information or materials or who have reviewed parts of the manuscript include Dr. Walter Biggs, Mr. Cal Bond, Dr. Phyllis Bradbury, Dr. Louis de Vos, Dr. Robin Leech, Dr. William Spooner, and Dr. Charles M. Williams. I am indebted to Carol Majors for several new drawings. I am also grateful for the assistance of my wife, Carol Lytle, for her editorial and secretarial assistance with this latest revision.

Laboratory Safety and Animal Care Procedures

Laboratory Safety

Biology laboratories are places for serious scientific work and study. Students and teachers must recognize that a number of potential safety hazards are present in all science laboratories. In a zoology laboratory, the principal safety hazards are electrical circuits, potentially dangerous chemicals, hot liquids and heat sources, broken glass, and sometimes infectious agents (pathogenic bacteria and viruses). Safety in the laboratory, as in other places, requires paying attention to potential hazards and observing appropriate safety measures. Nothing we do is without some degree of risk. People sometimes fall out of bed and injure themselves. Many people drown in swimming pools each year. Science laboratories do provide a safe environment when both students and teachers are aware of potential hazards and follow appropriate safe procedures.

The following list of safety rules is offered as a good start toward safe practice in lab. It is not a complete list of safety rules, however, and it is not a substitute for common sense. (Please note our first and last safety rules!)

Some Basic Rules of Safety for the Laboratory

1. Use common sense.
2. Avoid horseplay in the laboratory. A science laboratory is a place for work and contains many potential hazards for a careless person. Electrical circuits, potentially dangerous chemicals, and sometimes infectious agents (pathogenic bacteria and viruses) may be present in a zoology laboratory.
3. Never eat, drink, or smoke in the laboratory.
4. Always wash hands after handling chemicals or live or preserved animals.
5. Always wear shoes in the laboratory. Sandals are not appropriate.
6. Always be cautious when using electric hot plates and gas burners.
7. Use protective mittens or tongs to handle hot objects.
8. Wear safety goggles or other appropriate protective eye gear when engaged in performing or observing experiments or procedures that offer potential hazards to the eyes.
9. Be familiar with the properties and hazards associated with all chemicals used in the laboratory exercises. When in doubt about the hazards associated with any chemical, consult chemical material safety and data sheets (MSDS) distributed by all manufacturers. Check with your instructor for the appropriate data sheets.
10. Beware of electrical equipment with frayed or bare wires or faulty switches or plugs.
11. Be familiar with the operation and proper use of fire extinguishers, eyewash fountains, safety showers, and other safety equipment in the laboratory.
12. Clean all laboratory tables and other work surfaces after each use.
13. Keep scalpel blades sharp to avoid slipping and possible injury to yourself.
14. Make sure that all specimens you dissect are properly supported in a dissection pan or other appropriate surface. A specimen not properly supported could shift and lead to an injury from slippage of a scalpel or other dissecting instrument.
15. Work in a well-ventilated area when studying preserved specimens.
16. USE COMMON SENSE!

Handling and Care of Laboratory Animals

Experience in observing, handling, caring for, experimenting with, and dissecting animals is essential for the training of zoology students. Prospective employers and graduate and professional school admissions committees often cite the importance of such experience. Studies from textbooks, photographs, charts, models, and computer simulations may be good supplements, but they are not adequate substitutes for direct laboratory experience with living and preserved animals.

For this reason, it is important for every zoology student to learn the proper methods of handling and caring for animals. As a student of zoology and of nature, you should have a basic respect for all animal life and for the concerns of others about the humane treatment and conservation of animals.

The handling and treatment of vertebrate animals in teaching and research laboratories is regulated by federal law under the Animal Welfare Act of 1966 as amended in 1970, 1976, and 1985 plus additional regulations developed by the National Institutes of Health. Many states also have laws governing animal use. Invertebrate animals are not covered by these laws, but they should be treated with similar care and respect as living creatures. All teachers and researchers must be familiar with these laws and use good judgment and appropriate procedures when working with animals.

Always handle living animals with care and respect. With both vertebrate and invertebrate animals, take adequate precautions to avoid causing unnecessary stress or discomfort from your handling or experimenting. Any animals kept in the laboratory should have a clean and appropriate environment, adequate ventilation, food, and water, and regular care. At the conclusion of your studies, animals should either be disposed of in a humane manner or returned to a permanent animal care or culture facility, as directed by your instructor.

Further information about the care of animals in the laboratory and regulations governing animal care can be found in the following references.

National Research Council, 1988. *Use of Laboratory Animals in Biomedical and Behavioral Research,* 168 pp. ISBN 0-309-03796-4

National Institutes of Health, 1985. *Guide for the Care and Use of Laboratory Animals,* 83 pp. NIH Publication No. 86-23.

Orlans, F. Barbara, 1977. *Animal Care From Protozoa to Small Animals.* 374 pp. Addison-Wesley, Reading, Mass. ISBN 0-201-05484-1

Safety Precautions When Using Preserved Animals

The chemicals used to preserve animals and parts of animals can be toxic, flammable and/or dangerous if used improperly or under improper conditions. Ethanol, isopropanol, formaldehyde, phenol, and ethylene glycol are commonly used preservatives. The following information, supplied through the courtesy of the Carolina Biological Supply Company, provides some excellent safety guidelines to follow when handling and dissecting preserved animal specimens.

All teachers are responsible for implementing proper safety procedures when using potentially hazardous chemicals and for communicating appropriate information about these materials to their students in accordance with applicable federal, state, and local regulations.

To achieve the necessary level of safety in the laboratory, the instructor should be familiar with the chemicals present and the precautions to be taken when using these chemicals.

Carolina provides specimens preserved in alcohol, formaldehyde, and *Carosafe®* (contains ethylene glycol). We recommend these safety tips be followed whenever preserved specimens or chemicals are used:

1. Wear safety glasses at all times.
2. Do not wear contact lenses.
3. Wear appropriate gloves and lab coat.
4. Work only in a well-ventilated area.
5. Prohibit eating, drinking, and smoking in the work area.
6. In the event of contact, wash skin with soap and water; flush eyes with water.

Special chemicals may present hazards which require additional precautions to those listed above.

Formaldehyde should always be used in a well-ventilated area to prevent irritation to the eyes, skin, or respiratory tract. The use of goggles lessens eye irritation from formaldehyde vapors. If direct contact to eyes or skin occurs, wash thoroughly with water. Smoking should not be allowed.

Isopropanol is very flammable, so avoid sparks, open flames, and heat. Direct contact with isopropanol should be avoided through use of safety glasses, gloves, and lab coats. Wash thoroughly with water if direct contact to eyes or skin occurs.

Carosafe® is the safest of the three preservatives, but safety glasses should be worn to avoid direct contact with the eyes. Wash eyes thoroughly with water if contact occurs.

When working with preserved materials, be careful with sharp objects such as pins, scalpels, and the spines and teeth of specimens. When using a scalpel, we recommend cutting away from oneself and keeping fingers out of the cutting path.

Comparative Safety Data of Preservatives

	Formaldehyde	Isopropanol	Ethylene Glycol
Physical Data			
Hazardous Components (OSHA)	Methanol (TWA 200 ppm) Formaldehyde (TWA 3 ppm)	Isopropanol (TWA 400 ppm)	Ethylene glycol (NR)
Flash Point	184° F (combustible)	53° F (flammable)	241° F
Explosion Limits (lower)	7%	2%	3.2%
Extinguishing Media	Alcohol foam, water fog, carbon dioxide, dry chemical	Alcohol foam, carbon dioxide, dry chemical	Water fog, carbon dioxide, dry chemical
Unusual Fire Hazard	Vapor heavier than air; may travel along ground to ignition source	None	None
Threshold Limit Value (NIOSH)	2 ppm (air)	air: 400 ppm (skin)	100 ppm (vapor)
Effects of Overexposure			
Eyes	Vapor causes severe irritation, redness, tearing, blurred vision. Liquid may cause severe or permanent damage	Direct contact may cause irritation	Direct contact may cause irritation
Skin (direct contact)	Mild irritation, dermatitis, possible sensitization	Mild irritation	Mild irritation (Reddening of skin after 24 hours of continuous exposure)
Inhalation	Irritation of respiratory tract, dyspnea, headache, bronchitis, pulmonary edema, gastroenteritis	Irritation of respiratory tract; headache, nausea; at high concentrations, narcosis	Reported irritant effects at extremely high (10,000 mg/m^3) concentrations of vapor
Oral	rat LD_{50} 800 mg/kg	rat LD_{50} 5840 mg/kg	rat LD_{50} 5840 mg/kg

TWA:	Time Weighted Average
NR:	Not Restricted
ppm:	Parts per million; air (unless otherwise specified)

1

Microscopy

Objectives

After completing the laboratory work in this chapter, you should be able to perform the following tasks:

1. Identify the main parts of a compound microscope and explain their function.
2. Define and explain focus, working distance, resolving power, and magnification.
3. Describe the proper use and care of both compound and stereoscopic microscopes.
4. Explain the difference between compound and stereoscopic microscopes and give examples of appropriate uses of each type.
5. Use both compound and stereoscopic light microscopes in the correct way.
6. Explain the operating principles of a phase contrast microscope and give examples of its use.
7. Describe the two main types of electron microscopes and give examples of their use.
8. Explain the importance of microscopes in biological studies.

The Compound Microscope

The **compound microscope** is one of the most important and useful tools of the zoologist. It is used to study cells and cell parts, the organization of tissues, the structure of bone, and the structure of developing embryos, among many other important applications. Since many of the exercises in this course will require the use of the compound microscope, it is important to review some aspects of its construction, use, and care.

A modern compound microscope is illustrated in figure 1.1. Since there are numerous makes and models of compound microscopes in use, the microscope assigned for your use may differ slightly from the one illustrated. The operating principles and procedures, however, will be similar to those outlined later. Your laboratory instructor will point out any important differences between your microscope and the one illustrated.

A microscope is an expensive precision instrument and must be handled with care. Always carry your microscope with both hands. Grasp the arm of the microscope firmly with one hand and support the base with the other hand. Place it gently on your desk with the arm facing you.

Materials List

Living specimens
  Artemia (brine shrimp) larvae
Prepared microscope slides
  Letter ''e''
  Frog blood
Audiovisual materials
  Wall charts showing parts of compound and stereoscopic microscopes

---

Parts of the Microscope

Identify the principal parts and controls of the microscope with the aid of figure 1.1. At the top of the microscope is the **eyepiece,** or ocular lens, which is inserted in an inclined **body tube.** Microscopes with one eyepiece and body tube are monocular microscopes; those equipped with two eyepieces and body tubes for simultaneous viewing with both eyes are binocular microscopes. Below the eyepiece is the **arm** attached to the **base.** Also attached to the arm is the movable **stage,** which holds a microscope slide or other object for viewing. **Stage (slide) clips** aid in holding the slide in position on the stage. Above the stage is the **revolving nosepiece** with two or more **objective lenses.** Within the stage is another lens system, the in-stage **condenser,** which serves to concentrate light rays from the built-in **illuminator.** On the base near the illuminator is the **light switch.** Instead of a built-in illuminator, some microscopes have a **substage mirror** to reflect light from an auxiliary light source.

Beneath the condenser, locate the **disc aperture diaphragm.** Rotating the disc diaphragm increases or decreases the amount of light on the specimen. More expensive microscopes often have an **iris diaphragm** with

eyepiece

body tube

revolving nosepiece

objective lens

stage clip

stage

diaphragm control

built-in illuminator

base

arm

course adjustment

fine adjustment

power cord

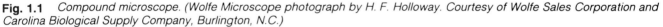

Fig. 1.1 *Compound microscope. (Wolfe Microscope photograph by H. F. Holloway. Courtesy of Wolfe Sales Corporation and Carolina Biological Supply Company, Burlington, N.C.)*

movable elements instead of a disc diaphragm. Raising and lowering the condenser also regulates the illumination of the specimen, although in most of your work in this course you will obtain satisfactory results by adjusting the condenser to the position that gives maximum illumination (usually near its uppermost position) and then making any further needed reductions in illumination with the disc diaphragm. This simplified method of controlling light does not produce the precise illumination required for advanced microscopy, but it produces results satisfactory for most routine purposes.

Accurate observation of a specimen requires positioning the objective lens at a specific distance from the specimen; this distance is determined by the specific construction of each objective lens and is called the **working distance** of the lens. When the objective lens is located at the proper working distance, the specimen will be in **focus.**

The working distance of the lens varies **inversely** with the magnification of the objective; low-power objectives have longer working distances and high-power objectives have shorter working distances.

Focusing the lens system is accomplished by mechanically changing the distance between the specimen and the objective lens. Coarse and fine **adjustment knobs** for this purpose are provided on the side of the arm near the base. On some modern microscopes both coarse and fine adjustment are controlled by a single knob with dual function, but most models have separate controls. How many focusing control knobs are there on your microscope?

When you observe a specimen through a compound microscope, several images are formed by the optical system of the microscope. Three of these images are of special importance—the **real image,** the **virtual image,** and the **retinal image.**

The **real image** is formed at a specific distance above the objective lens. What you actually see when you look through the objective lens is the **virtual image**, which appears both larger in size and farther away than the specimen on the stage. The **retinal image** is formed by the rays of light striking the retina of your eye. The virtual image can be helpful to you in estimating the size of an object viewed through a microscope as explained later.

Magnification

The principal purpose of a microscope is to magnify the image of an object. The **magnification** of an object is determined by the construction of the ocular and objective lenses of the microscope, and the total magnification is the product of the separate magnification of these two lenses.

Example:

10X ocular × 10X objective = 100X total magnification

Student microscopes used in introductory biology and zoology courses are commonly equipped with 10X ocular lenses, and both 10X and 43X objective lenses. These two lenses are mounted on a revolving nosepiece and are called the **low-power** and **high-power** objectives, respectively. Often, a 3.5X objective lens is also used on student microscopes; this is called a **scanning lens.** It is useful for viewing relatively large objects or for preliminary location of a specimen on the microscope slide. Occasionally a 90X or 100X **oil immersion lens** may be present for viewing very small objects like bacteria. Special instructions and precautions are needed for the use of oil immersion lenses.

Magnification is changed in microscopes with multiple objectives by rotating the revolving nosepiece until the desired objective is in position below the body tube of the microscope. A newer type of compound microscope (zoom lens type) uses a system of movable lenses to produce a variable magnification rather than a series of fixed magnifications as in a microscope equipped with a rotating nosepiece.

Observe the 10X objective on your microscope and note that it is also marked 16mm (or some similar value). This is the **working distance** of that lens; thus, this lens will be in focus at approximately 16mm above the surface of your specimen. What is the working distance of your 43X objective?

All the other parts of the microscope are accessory to the main purpose of magnification by the lenses. They consist mainly of mechanical devices to hold the specimen, to regulate the light necessary for clear vision, and to facilitate focusing.

Measuring the Size of an Object

The size of an object viewed in a compound microscope can be estimated in several ways. The simplest method involves the use of a small square of graph paper or a small plastic ruler of the type found in most dissecting kits to measure the size of the virtual image. The length of the virtual image is then divided by the theoretical magnification of the ocular lens times the magnification of the objective lens.

$$\text{size of object} = \frac{\text{length of virtual image}}{\text{theoretical magnification}}$$

This procedure gives only a rough approximation because the actual magnification of the ocular and objective lenses is slightly different from the stated theoretical magnification, and your estimate of the size of the virtual image is not precise.

Instructions for estimating the size of an object with this method are given in the following exercises on the use of the compound microscope at the end of this discussion of the compound microscope.

More precise measurement of objects with a compound microscope requires the use of an ocular micrometer—a small glass disc etched with a measuring scale mounted inside one eyepiece of the microscope. An ocular micrometer is calibrated with a stage micrometer and provides the most accurate method for measuring microscopic objects. If your microscope is equipped with an ocular micrometer, ask your instructor for assistance in calibrating it so you can measure objects with your microscope.

Resolving Power

While magnification is the increase in size of the image of an object, the ability of a lens or a microscope to reveal the fine detail of a specimen is called its **resolving power.** In microscopy, an increase in the apparent size of a specimen is not always accompanied by an increase in the clarity of detail within the specimen. Beyond certain limits, further magnification simply makes the apparent image of a specimen become progressively fuzzy or indistinct as it becomes larger. The resolving power of a microscope depends largely upon the design and quality of its objective lenses, and it is really resolving power rather than magnification that limits the **useful** magnification of a compound microscope. Magnifications beyond approximately 2,000 diameters require an entirely different microscopic system.

Technically, the resolving power is defined as the shortest distance between two points that can be visually distinguished as separate points.

Illumination

Proper illumination of the specimen is an extremely important matter, since improper lighting of the specimen can produce poor images, inaccurate observations, and/or unnecessary eyestrain.

Most modern student compound microscopes are equipped with in-base illuminators. If you have such a microscope, you can adjust the amount of light reaching the objective lens by rotating the disc (or iris) diaphragm. Experiment with the diaphragm control and observe the changes in amount of light as the diameter of the diaphragm opening changes. You will find that the best illumination for viewing specimens is obtained with the smallest diaphragm opening that lights the entire microscopic field.

Sometimes illumination for a compound microscope is provided by a separate lamp and a substage mirror. If you are assigned a microscope of this type, examine the two surfaces of the mirror—one surface is flat, and one surface is concave. The flat surface of the mirror is the one used most often in normal laboratory situations. When the mirror is properly adjusted, light is reflected by the mirror and passes through the condenser, the specimen, and the lenses to your eye. The concave (curved) surface of the mirror is used less frequently and only in certain situations where there is a need to concentrate the light rays. Adjust your mirror to reflect the maximum light on the specimen, and reduce the illumination as necessary by closing the iris diaphragm.

Illumination is also controlled by the substage condenser. For routine work in an introductory laboratory, the condenser should not be used to adjust illumination, but should be set at or near its uppermost position, and the light intensity adjusted by means of the diaphragm.

Focusing

To obtain a clear image of a specimen, you must carefully adjust the distance between the lenses and the specimen. This adjustment of the spatial relationships between the lenses and the specimen is called **focusing.** When a clear image of the specimen can be seen through the ocular lens, the specimen is said to be **in focus.** At this adjustment of the lens system, the specimen is located precisely at the working distance of the objective lens.

■ Locate once again the coarse adjustment knob on your microscope and turn the control slightly in one direction. Observe that movement of the knob increases or decreases the distance between the stage and the objective lens.

If you have a microscope with separate coarse and fine adjustment knobs, carefully try each control to see the difference in their actions. You will find that it takes several turns of the fine adjustment control to equal the effect of a single turn of the coarse adjustment control.

While testing the action of the focusing controls, watch the objective lens and the specimen stage. Which part moves and which part remains stationary?

Caution: Never allow the objective lens to touch a slide or the condenser lens, which may protrude through the hole in the specimen stage. Anything coming in contact with the objective lens presents a danger of scratching, cracking, or otherwise damaging this critical and expensive part of the microscope. Always remember that this expensive scientific instrument is assigned to you for your use and safekeeping. Be certain that you use it properly and carefully, and that you keep it in good condition. Promptly notify your laboratory instructor of any malfunctions or if you have any difficulty in the use of your microscope.

Procedure for Use of the Microscope

Good microscopy requires the adoption of work habits that facilitate observation, minimize fatigue and eyestrain, and protect the equipment from damage. The steps outlined below should be followed each time you use the compound microscope.

1. Place the microscope directly in front of you on the laboratory table. Remember to carry the microscope by using both hands after removing it from the storage cabinet. Clean the objective and ocular lenses by wiping gently with a clean sheet of lens paper; never use anything but lens paper for cleaning the lenses.

2. Rotate the revolving nosepiece until the low-power (10X) objective lens clicks in place directly over the center of the condenser. While viewing the objective lens from one side of the microscope tube, **carefully** adjust the distance between the lens and the specimen stage, using the coarse adjustment control, to approximately one-half inch.

3. Select a microscope slide provided by your laboratory instructor and examine it to locate the position of the specimen on the slide. Then place the slide on the stage, with the specimen centered over the condenser lens. Make certain that the cover glass and specimen are on top of the slide.

4. Open the disc (or iris) diaphragm fully and turn on your microscope lamp. If your microscope is equipped with a substage mirror and an auxiliary lamp, adjust the positions of the lamp and the mirror until you obtain an evenly lighted, circular microscope field. Reduce the light on the specimen as necessary for clear observation by adjusting the diaphragm.

5. While observing the objective lens and the microscope slide from the side again, carefully readjust the distance between them to about one-eighth inch. DO NOT LOOK THROUGH THE OCULAR WHILE PERFORMING THIS STEP.

6. Now look through the ocular and slowly **increase the distance between the objective lens and the specimen** until the specimen comes into focus. This may involve either raising the objective lens or lowering the specimen stage, depending upon the design of your particular microscope. **Never** focus in the opposite direction—by decreasing the distance between the objective lens and the specimen—**while looking**

through the ocular. Further, center the specimen in your field of view as necessary, and bring the specimen into sharper focus with the fine focusing control.

7. Rotate the revolving nosepiece until the high-power objective lens (43X) clicks into position directly over the condenser lens. The objective lenses on most modern microscopes are factory installed and adjusted so only minor changes in focusing are necessary when magnification is changed. Such lenses are **parfocal**—that is, their planes of focus are identical, or nearly so, although their working distances are different. You should find it necessary to make only a slight adjustment with the fine focusing control in order to achieve good focus after changing from the 10X objective to the 43X objective. **Use only the fine focus control** while looking through the ocular with the high-power objective in position. You will also find it necessary to open the iris diaphragm slightly when you switch from low to high power. Why?

Returning the Microscope after Use

1. Rotate the revolving nosepiece to place the **low-power objective** (or scanning lens if your microscope has one) into position over the condenser lens.

2. Remove the microscope slide from the specimen stage and return it to its proper box or tray. If you have been using wet mounts, clean the specimen stage with a **clean cloth** or **cleaning tissues** as provided by your laboratory instructor.

3. Clean the objective and ocular lenses of the microscope with **lens paper.**

4. If your microscope has an in-base illuminator, disconnect the power cord, and carefully wind the cord as directed by your instructor.

5. Return the microscope to its storage cabinet. Remember to use both hands in carrying the microscope. Check once again to make certain that the low-power objective lens (or scanning lens) is in position and that you have not left a slide on the stage.

Special Precautions

1. **Never** focus down (raise the stage or lower the objective, depending on the type of your microscope) while looking through the microscope.

2. **Always** locate the specimen under low power before switching to high power.

3. **Never** turn your microscope upside down or lay it on its side. The ocular might fall out and could be damaged.

4. **Always** keep the microscope clean and dry. Use only lens paper to clean the lenses.

5. **Never** use the coarse focusing control when the high-power lens is in position. Focus only with your fine focusing control when using high power.

6. **Always** try to relax and keep both eyes open when using the microscope. This helps to prevent undue eyestrain. With a little practice, you can learn to concentrate on the specimen and to disregard the image received by the other eye if you are using a monocular microscope.

Exercises Using the Compound Microscope

1. Obtain a microscope slide with the letter "e" on it. Place the slide on the stage of your compound microscope and observe under low power. What is the orientation of the letter "e" as you view it through the compound microscope? Why does it look different from a letter "e" printed on this page?

   Move the microscope slide slightly to the left while you look through the eyepiece. Which way does the letter "e" seem to move?

   Move the slide slightly away from you. Which direction does the letter "e" move this time? Change objective lenses to high power by rotating the nosepiece or by changing the zoom adjustment if you have a zoom-type microscope. Observe the letter "e" under high power. How much of the letter "e" can you see under high power? Note that the letter seems larger and that your field of view is smaller at high power than at low power.

2. Obtain a slide of frog blood and study it first under low power and later under high power. What is the shape of the cells? What structures can you observe in the cell? Draw a frog blood cell on the page for Notes and Sketches at the end of this chapter. Label the cell parts you recognize. You will learn more about cells and their parts in the next chapter.

3. Mount a small square of millimeter-square graph paper on a clean microscope slide under a coverslip. Focus and observe under low power. Carefully move the slide to center one millimeter square in your microscope field with two sides of the square parallel to the edge of the microscope stage. Place a small plastic metric ruler on the right edge of your microscope stage so that you can see the ruler with one eye while you view the slide with the graph paper with your other eye.

   Keep both eyes open and focus your eyes on both images with the square of graph paper in one eye and the scale of the ruler with the other eye. It will seem strange at first to look at two different images simultaneously, but if you relax and continue to observe both objects, they will soon become superimposed in your vision.

When you see the two superimposed images, measure the apparent size of the millimeter square as it appears on the superimposed ruler. Since the lines surrounding the square appear relatively thick, be sure that you measure from the upper edge of one line to the upper edge of the next line to compensate for the thickness of the lines in your measurement.

Record your measurement of the millimeter square in your Notes and Sketches. Also, sketch the superimposed image you saw with both eyes, the millimeter square and the ruler on the microscope stage.

You can calculate the magnification under low power by dividing the apparent size of the millimeter square by its actual size (1 mm).

Example:

$$\frac{103 \text{ mm (observed size)}}{1 \text{ mm (known size)}} = 103\text{X magnification}$$

You can also use this procedure to calculate the approximate size of a specimen observed under your microscope. If you observe a cell under the microscope that appears to be 25 mm in diameter, simply divide by the calculated magnification to determine the actual size of the cell.

Example:

$$\frac{25 \text{ mm (observed size)}}{103 \text{ (magnification)}} = .24 \text{ mm}$$

Calculation of the size of an object viewed under high power is a bit more complicated, but follows the same principle. First, place the ruler on the stage and measure the apparent diameter of the microscope field at high power. Next, place the slide with millimeter square graph paper on it on the stage and orient the slide again so the sides of the square are parallel to the ruler. Switch to high power and move the millimeter square so that one corner is at one edge of the field. Locate some distinctive landmark on the baseline of the square, such as a bulge in the line or a large fiber, at the opposite edge of the field. Carefully move the slide to place your landmark directly across the field and find a new landmark on the opposite side of the field again. Continue this procedure until you determine how many microscope fields at high power it takes to traverse the one millimeter square.

Multiply the apparent field diameter as measured with the ruler by the number of diameters required to traverse a one millimeter square to obtain the magnification of your microscope at high power.

Example:

apparent diameter of field $\times$ number of fields to traverse 1 mm = magnification

or

175 mm $\times$ 2.4 field diameters = 420X

In a like fashion, you can estimate the actual size of an object by dividing its apparent size (measure the virtual image with a ruler) by the magnification under high power.

Example:

$$\frac{\text{apparent size}}{\text{magnification}} = \frac{160 \text{ mm}}{420\text{X}} = 0.38 \text{ mm}$$

The values obtained by these procedures are only approximate but provide size and magnifications useful for most purposes. More precise measurement of microscopic images require more complicated and expensive equipment.

The Stereoscopic Microscope

The **stereoscopic microscope,** also frequently called the dissecting microscope or stereomicroscope, is another common and extremely useful laboratory instrument (see figure 1.2). Stereoscopic microscopes are useful for viewing objects such as small insects, frog eggs, or large protozoa at low magnifications.

The image seen through a stereoscopic microscope is not inverted, as is the image seen through a compound microscope. Also, the internal configuration of lenses and prisms provides dual light paths and thus produces a stereoscopic or three-dimensional image. The effective magnification of such a lens system is more limited than is that of compound microscopes. Most stereomicroscopes provide magnifications in the range of 5–50X, although useful magnification of up to 100–200X can be obtained in the best quality research stereomicroscopes.

Two recent designs of stereomicroscopes now in common use in teaching, as well as in research laboratories, utilize different mechanical and optical systems to achieve changes of magnification. One of these types of stereomicroscopes employs a fixed-position ocular, objective paired lenses, and an internal rotating drum on which several paired prisms are mounted. Two large knobs located on either side of the microscope head serve to rotate the internal drum and thus to change magnification. Focusing is accomplished by raising or lowering the objective, as in a compound microscope. The **focusing control knobs** are located on the sides of the microscope arm.

The other newer type of stereomicroscope utilizes a zoom-type lens system and provides a continuously variable magnification by the rotation of a cylinder located above the objective lens (see figure 1.2). Focusing is accomplished in this type of stereomicroscope by means of two lateral focusing control knobs as in the previously discussed types of stereomicroscopes.

Stereomicroscopes may be equipped with either an opaque or a transparent glass stage plate or disc. When equipped with a transparent stage plate and a substage beneath the regular stage, a stereomicroscope can be used to view objects in transmitted as well as in reflected light. This is often a very useful feature for biological studies.

eyepiece

eyepiece adjustment

binocular head

body

magnification control

adjustment knob

objective lens

lamp for reflected light

glass stage plate for transmitted light

lamp switches

power cord

base

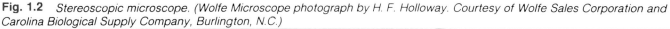

Fig. 1.2 *Stereoscopic microscope. (Wolfe Microscope photograph by H. F. Holloway. Courtesy of Wolfe Sales Corporation and Carolina Biological Supply Company, Burlington, N.C.)*

Exercises in the Use of the Stereomicroscope

1. Remove your stereomicroscope from its storage cabinet and determine which type has been provided for your use in this course. Are there other types of stereomicroscopes present in the laboratory? Compare them with the type you have.

2. Place your stereoscopic microscope directly in front of you on the table and focus a concentrated light on the center of the stage. Select a microscope slide with a relatively large specimen, such as a fluke, a tapeworm, an insect wing, or other suitable specimen, and examine it through the microscope.

3. Experiment with several such slides as provided by your laboratory instructor until you become familiar with the use of your stereomicroscope. Try various adjustments of your light on the specimen, and try both transmitted and reflected light on specimens that are transparent or translucent if your microscope is provided with a substage and a substage mirror. If you do not have a substage, you may be able to achieve a similar effect by illuminating the specimen at right angles to your plane of viewing. Place your lamp close to the tabletop and focus the light beam from the side as sharply as possible on

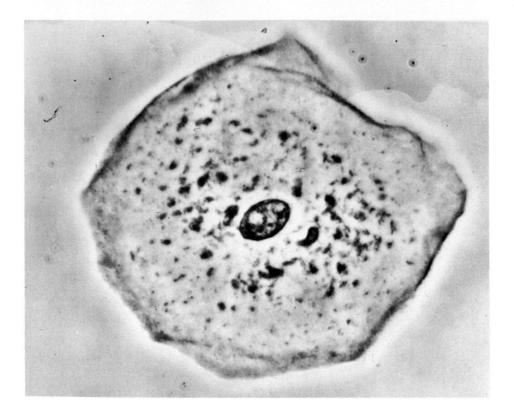

Fig. 1.3 *Phase contrast photograph of a living cell from human cheek epithelium. (Courtesy of Carolina Biological Supply Company, Burlington, N.C.)*

your specimen. Can you observe any structural details not discernible when the light comes from above or at a higher angle?

4. Place a few newly hatched larvae of the brine shrimp *Artemia,* or of a similar small living animal, in a watch glass and examine them in both direct and transmitted light. Make a simple outline drawing showing the major structures that you see.

5. When you have completed your study with the stereoscopic microscope, wipe the stage clean of any spills, using a clean cloth or cleaning tissue, and return the microscope to its storage cabinet.

Other Types of Microscopes

Compound and stereoscopic microscopes are only two of the several different types of microscopes that have been developed by scientists in their continuing efforts to observe small objects more closely and in greater detail. Two other kinds of microscopes of particular importance in zoology are the phase contrast microscope and the electron microscope. In their design, both types of microscopes employ physical principles that differ significantly from those employed in ordinary light microscopes.

Phase Contrast Microscopy

Phase contrast microscopy permits the direct examination of transparent, unstained materials, including many kinds of living cells and tissues. Basically, this type of microscopy depends upon a special method of illumination which increases the contrast in an unstained specimen resulting from slight variations in the thickness and refractive index of its parts. The light passing through the specimen is manipulated in the lens system in such a way that these minor physical differences within the specimen are transformed into varying degrees of brightness and darkness. Thus, a living cell viewed through a **phase contrast microscope** has the appearance of being stained, although no chemicals that might alter its structure or kill it have been added to the cell. Figure 1.3 illustrates the type of image obtained in this kind of microscopy.

Fig. 1.4 *Transmission electron microscope equipped with a special X-ray analyzer on the right. (Photograph courtesy of Philips Electronic Instruments, Inc.)*

Electron Microscopy

Electron microscopy has become a very valuable research technique in zoology in the last 30 to 40 years. In electron microscopy, a specimen is irradiated with a concentrated **beam of electrons** rather than with rays of visible light, as in a light microscope. The two main electron microscopes commonly used by zoological researchers are **transmission electron microscopes** and **scanning electron microscopes.**

Transmission electron microscopes (figure 1.4) are capable of much greater resolving power and, therefore, much higher magnifications than are ordinary light microscopes. Direct magnifications up to 200,000 diameters are commonly achieved with electron microscopes compared to a maximum of about 2,000 diameters with a compound light microscope. Photographic enlargements yield final magnifications greater than 1,000,000 diameters in many electron micrographs. Figure 1.5 is an example of a photograph taken with a transmission electron microscope.

The **scanning electron microscope** (figure 1.6) is an even newer research tool and has had many important applications in biological research during the past 20 years. The scanning electron microscope differs in principle from both light and transmission electron microscopes. This kind of electron microscope has proved to be especially useful in providing three-dimensional images of small objects, information about chemical composition, electrical properties, and structural details of the surface of specimens.

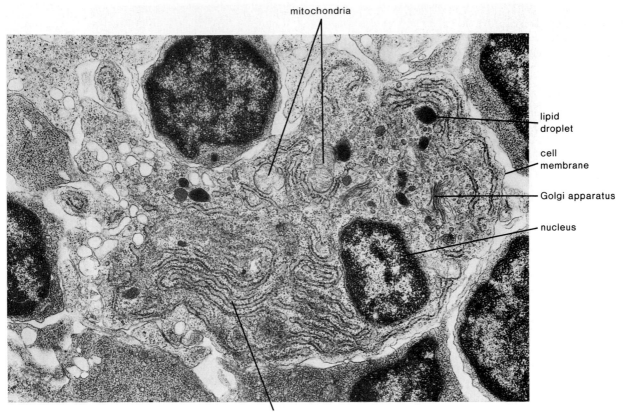

mitochondria

lipid
droplet

cell
membrane

Golgi apparatus

nucleus

endoplasmic reticulum

Fig. 1.5 *Transmission electron micrograph of a plasma cell. (Photograph by Kenneth E. Muse.)*

Fig. 1.6 *Scanning electron microscope. (Photograph courtesy of Philips Electronic Instruments, Inc.)*

Fig. 1.7 *(left) Scanning electron micrograph of an ant. Magnification 300×. (right) Enlarged view of the foreleg joint from the circled area of the illustration at left. Magnification 600×. (Photographs by Kenneth E. Muse.)*

The scanning electron microscope has a great depth of field (7 to 10 times that of a light microscope at comparable magnifications), making possible photographs of excellent three-dimensional quality (see figure 1.7).

Key Terms

Compound microscope   a type of light microscope with two separate lens systems, an eyepiece and an objective lens, which together serve to magnify the image of an object. Provides magnification to about 1,000–2,000 diameters.

Magnification   the ratio of the apparent size to the actual size of an object when viewed through a microscope.

Parfocal lenses   objective lenses constructed and mounted so that their focal planes are approximately the same; a specimen remains in focus when you change from one parfocal lens to another.

Phase contrast microscope   a special type of light microscope that permits the observation of thin, unstained materials. Special lenses and illumination techniques increase the contrast in an unstained specimen due to variations in the refractive indices of its parts.

Resolving power   the ability of a microscope to reveal fine detail in a specimen. More precisely, resolving power is defined as the shortest distance between two points that allows them to be distinguished as separate points.

Scanning electron microscope   a type of electron microscope that provides three-dimensional images of very small objects. A concentrated beam of electrons is focused and moved along the surface of a specimen and induces the emission of secondary electrons from the specimen. These secondary electrons produce a magnified image of the specimen on a cathode-ray tube.

Stereoscopic microscope   a type of microscope with two separate optical paths that provide a magnified three-dimensional image. Provides useful magnifications to about 100–200 diameters.

Transmission electron microscope   a type of electron microscope used to view specially prepared thin specimens at magnifications to about 200,000 diameters. A concentrated beam of electrons passes through the specimen and produces a pattern of light and dark areas on a phosphorescent screen because of the differential passage of electrons through portions of the specimen. Darker areas represent areas of greater electron density within the specimen.

Working distance   the distance from the front of the objective lens of a compound microscope to the top of the specimen.

Notes and Sketches

2

Animal Cells and Tissues

Objectives

After completing the laboratory work in this chapter, you should be able to perform the following tasks:

1. Describe the Cell Theory and explain its importance in zoology.
2. Describe the principal organelles of a typical animal cell, such as a starfish egg, that are visible in a light microscope.
3. List six organelles seen in a transmission electron micrograph of a generalized (typical) animal cell not generally visible with a compound microscope.
4. Distinguish between a cell, a tissue, an organ, and an organ system.
5. List five types of animal tissues and give examples of each. Identify typical examples of each in microscope slides.
6. Describe the histological structure of a compact bone and explain the role of a Haversian canal, lacuna, canaliculi, and lamellae.
7. Distinguish between smooth muscle, cardiac muscle, and striated muscle from microscopic slides or photographs.
8. Describe the structure of a vertebrate neuron and give the function of each main part.
9. Describe the composition of human blood and give the functions of erythrocytes, leucocytes, and platelets.
10. Identify erythrocytes, leucocytes, and blood platelets in microscopic preparations.

The Cell Theory

Cells are the fundamental structural and functional units of virtually all living organisms. The body of an animal typically is made up of many different kinds of cells that are organized into tissues, organs, and organ systems that carry out certain essential functions. Thus, a knowledge of cell structure and function is essential for the proper understanding of reproduction, growth, heredity, and all other normal and abnormal animal functions.

The importance of cells is summarized in a statement called the Cell Theory, which is one of the most important unifying concepts in biology. This theory was developed in the nineteenth century, when naturalists were experimenting with their latest technology—the compound microscope. The Cell Theory is generally attributed to two German scientists, botanist Matthias Schleiden and zoologist Theodor Schwann, who published their ideas in 1838–1839; but other scientists also have contributed to the modern version of the theory.

The main points of the Cell Theory can be summarized as follows:

1. All organisms are composed of cells.
2. All cells come from other cells.
3. All vital functions of an organism occur within cells.
4. Cells contain the hereditary information necessary for regulating cell functions and for transmitting information to the next generation of cells.

Prepared microscope slides

Starfish eggs or sea urchin eggs
Cuboidal epithelium (rabbit kidney)
Columnar epithelium (rabbit kidney or amphibian intestine)
Stratified epithelium (human skin)
Loose connective tissue
Hyaline cartilage
Human bone, cross section
Smooth muscle
Striated muscle
Neurons (smear from spinal cord of cow)
Blood film (human), Wright stain
Ciliated epithelium (Demonstration)
Adipose tissue (Demonstration)
Cardiac muscle (Demonstration)
Amphibian blood (Demonstration)

Chemicals

Wright's blood stain
Distilled water

Miscellaneous supplies

Clean microscope slides
Coverslips
Fresh cartilage (from frog sternum, ends of long bone, etc.)
Cross sections of long bones from pig, cow, or other mammal

Basic Cell Structure

Our knowledge of cell structure has increased dramatically in the last few decades because of the availability of electron microscopes as described in chapter 1. Figure 2.1 illustrates the appearance of a relatively simple cell seen in an ordinary compound light microscope of the type usually available in general zoology laboratories. Figure 2.2 is a diagram illustrating the parts of an animal cell seen in transmission electron micrographs. Note the obvious difference in structural detail. We now know that animal cells are composed of many kinds of organelles, the distinctive parts of cells that carry out specific functions. Several types of cellular organelles can be visualized only in electron micrographs because they are too small to be seen with a compound microscope.

Since electron microscopy is a very complex process, in this exercise we will concentrate on those aspects of cell and tissue structure observable with a compound microscope and compare our findings with information available from more sophisticated techniques and instruments.

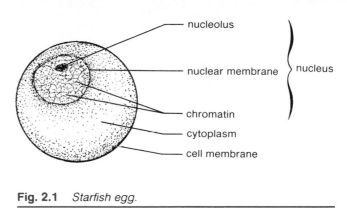

Fig. 2.1 *Starfish egg.*

Good examples of undifferentiated animal cells are provided by the unfertilized eggs of many animals. For our introductory study of cells, we will use the unfertilized eggs of the starfish. Sea urchin eggs are very similar to starfish eggs and serve equally well. Prepared microscope slides containing many stained starfish eggs will be provided for your study.

■ Obtain a prepared microscope slide and, under low power on your compound microscope, observe the numerous starfish eggs. Select a well-stained cell similar to that illustrated in figure 2.1 and move the microscope slide to center the cell in your field of view. Rotate the high-power objective into position, regulate the light as needed, and readjust the focus. Observe the two well-differentiated parts of the **cell,** the central **nucleus** and the surrounding **cytoplasm.** Externally, the cytoplasm is bounded by a thin **cell membrane.** Although the cell membrane of animal cells is seen only as a thin outer boundary of the cell, it plays a very important role in the functions of the cell. The special properties of the cell membrane control the passage of materials into and out of the cell. Animal cells lack the thickened cellulose cell walls outside of the cell membrane that are usually found in plant cells.

■ In your specimen, identify the spherical **nucleus,** bounded by the **nuclear membrane** and containing numerous darkly staining masses of **chromatin** material. Also within the nucleus of the starfish egg, find the darkly staining **nucleolus.** On table 2.1, list the cell organelles you were able to observe with a compound microscope. In table 2.2, make a similar list of all the additional cellular organelles that you find

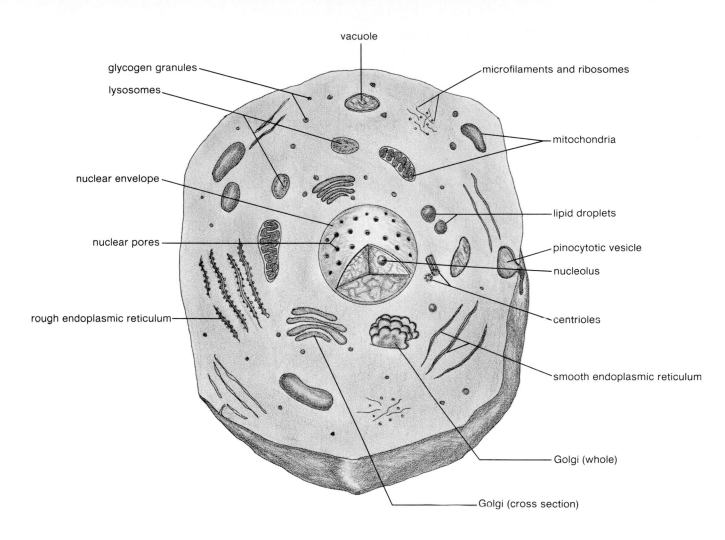

Fig. 2.2 *Generalized animal cell, as seen with the aid of a transmission electron microscope.*

Table 2.1.   Cellular Organelles Observed with Light Microscope.

1. _____
2. _____
3. _____
4. _____
5. _____
6. _____

Table 2.2.   Cellular Organelles Commonly Observable in Transmission Electron Micrographs but NOT Seen with Light Microscope.

1. _____
2. _____
3. _____
4. _____
5. _____
6. _____
7. _____
8. _____
9. _____
10. _____

in the diagram of an animal cell in figure 2.2 and in the transmission electron micrographs in figure 1.6, but were unable to observe with your light microscope. What conclusions can you make from a comparison of the two lists?

Animal Tissues

The bodies of most kinds of animals are multicellular, and in most multicellular organisms the cells are organized into definite tissues. A **tissue** can be defined simply as an aggregate of cells performing a similar function or functions. Tissues, in turn, are organized into **organs,** which consist of one or more tissues grouped into a structural and functional unit. Organs with related functions work together as **organ systems** in many types of animals. An **organism** consists of several integrated organ systems.

Animal tissues are commonly divided into four principal types: (1) **epithelial tissue,** (2) **connective tissue,** (3) **muscular tissue,** and (4) **nervous tissue.** Two other tissue types are also sometimes distinguished: (1) **vascular tissue** or blood, which is really a specialized type of connective tissue, and (2) **reproductive tissue,** which includes the specialized reproductive cells or gametes: eggs and sperm (see chapter 4).

Epithelial Tissue

In epithelial tissues, the cells are closely spaced with small amounts of intercellular substance and are arranged in thin layers or sheets. They form the covering of the outer surface of the body, line the internal cavities and ducts, and form glands. There are several kinds of epithelial tissue, each distinguished by the shape of the cells and their arrangement.

Squamous epithelial tissue consists of thin, flattened cells arranged in one or several stratified layers. They line cavities such as the mouth, esophagus, vagina, and the coelom where the exchange of materials through absorption and diffusion of gases and other materials is especially important to the underlying cells. The cells lining your mouth cavity provide a good example of this tissue type (see figure 1.3).

■ Take a clean, blunt toothpick and gently scrape the inside of your cheek to remove a few of the surface epithelial cells. When properly done, this operation is painless and bloodless. Touch the end of the toothpick to the center of a clean microscope slide and roll the toothpick carefully back and forth to transfer some of the epithelial cells to the slide. Add a drop of 1% methylene blue solution and a coverslip. Observe your preparation under a compound microscope. Draw some of the squamous epithelial cells from the lining of your mouth in the space provided

Fig. 2.3 *Student drawing of squamous epithelial cells from human skin.*

in figure 2.3. Other types of epithelial tissues include **squamous epithelium, cuboidal epithelium, columnar epithelium,** and **stratified squamous epithelium** (figure 2.4).

Cuboidal epithelium is another type of epithelial tissue found lining several types of ducts such as kidney tubules, salivary glands, and the secretory follicles of thyroid glands. These cells are sturdier than squamous epithelium in order to withstand the frequent abrasions from the passage of materials.

■ Examine a slide of cuboidal epithelium (figure 2.5) from the kidney tubules of a rabbit or other appropriate tissue. Note that the height and width of the single layer of cuboidal cells are about equal. A **basement membrane** lies at the base of the cells. Note the prominent nucleus.

Columnar epithelium (figure 2.6) consists of a single layer of tall, closely packed cells.

■ Obtain a slide with columnar epithelium from the intestine of a frog, rabbit kidney, or other appropriate tissue, and identify the outer **brush border,** the **nucleus,** and the **basement membrane.** Columnar epithelium from the intestine also often exhibits secretory **goblet cells.** Some columnar epithelial tissues, such as those lining the oral cavity of the frog and the human trachea, bear **cilia** on their distal (outer) margins.

Chapter 2

Fig. 2.4 *Common types of epithelium.*

Fig. 2.5 *Cuboidal epithelium. (Courtesy of Carolina Biological Supply Company, Burlington, N.C.)*

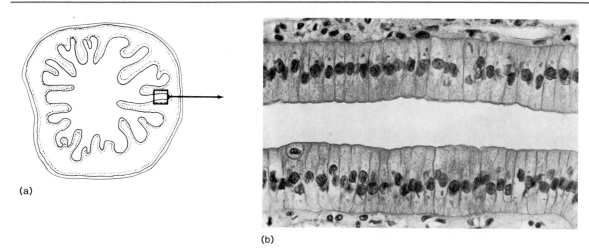

(a)

(b)

Fig. 2.6 *(a.) Frog intestine cross section; (b.) Columnar epithelium. (b. Courtesy of Carolina Biological Supply Company, Burlington, N.C.)*

epidermis

dermis

subcutaneous layer

hair shaft

stratified
epithelium

erector
muscle

sebaceous
gland

loose
connective
tissue

sweat gland

adipose tissue

nerve   hair follicle

Fig. 2.7 *Human skin, cross section.*

Stratified epithelium (figure 2.7) makes up the outer part of the skin or integument of all vertebrates. This type of epithelial tissue consists of several layers of similar cells that become progressively flattened as they rise to the surface. Stratified squamous epithelium is adapted for surfaces with frequent abrasion such as the human skin (figure 2.7) where the surface layers are constantly being worn and sloughed off.

■ Obtain a slide of human skin and observe the complex structure of the multilayered integument, or outer covering of the human body.

Connective Tissue

All connective tissues exhibit relatively large amounts of nonliving, intercellular substance produced by the living cells. This substance provides support and protection for the animal body. You will study three examples of connective tissue in this exercise: **loose connective tissue, cartilage,** and **bone.**

Loose Connective Tissue

Loose connective tissue (figure 2.8) consists of scattered cells surrounded by a clear, jellylike **ground substance** and two types of fibers: **elastic fibers** and **nonelastic** (collagenous) **fibers.** Loose connective tissue is often called areolar tissue. It is found connecting the various tissues of the body in humans and other vertebrate animals. Nerves, blood vessels, and cells lie embedded in loose connective tissue.

■ Obtain a slide of loose connective tissue and observe the cells, the elastic and nonelastic fibers, and the apparently open spaces where the ground substance has

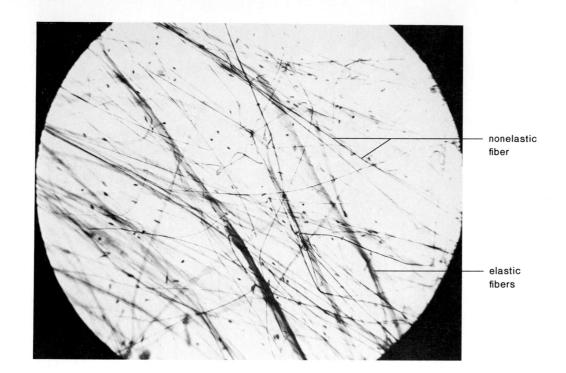

Fig. 2.8 *Loose connective tissue. (Courtesy of Carolina Biological Supply Company, Burlington, N.C.)*

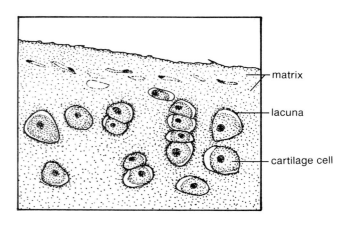

Fig. 2.10 *Cartilage tissue.*

Cartilage

Cartilage (figure 2.10) consists of a firm but elastic **matrix** secreted by numerous cartilage cells embedded within the matrix. **Hyaline cartilage** is found at the ends of long bones. Cartilage cells are found in open spaces or **lacunae** (singular: lacuna) scattered within the matrix (chondrin) which they secrete. Most cartilage cells are isolated in a lacuna separated by a relatively large amount of intercellular material. Occasionally, however, you can find two recently divided cells within a single lacuna. Following their division, the two new cartilage cells begin to secrete more chondrin, and each new cell creates its own new lacuna. This is how cartilage tissue grows and how it can be repaired after injury.

Fig. 2.9 *Student drawing of elastic and nonelastic fibers.*

been dissolved during the preparation of the slide. Identify the thin, wavy, **nonelastic fibers** and the thicker **elastic fibers.** Other types of connective tissue include **adipose** (fat) **tissue, cartilage,** and **bone.**

Draw examples of elastic and nonelastic fibers in figure 2.9 and label each type.

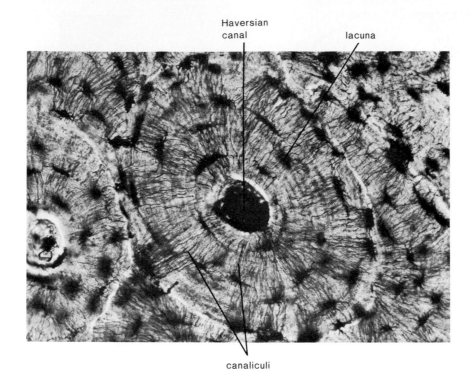

Haversian canal

lacuna

canaliculi

Fig. 2.11 *Bone tissue. (Courtesy of Carolina Biological Supply Company, Burlington, N.C.)*

■ Study a microscope slide of hyaline cartilage and observe the scattered cells, lacunae, and chondrin matrix. Can you find two recently divided cartilage cells in your microscope slide? How would you identify them as recently divided cells?

Bone

Bone plays an important role in the mechanical support of the body, and also in protecting vital parts from injury. The skull and the vertebral column of humans, for example, play dual roles in the support and protection of the brain and the spinal cord as well as support for the main part of the body. The heart and the lungs are also protected by the bony framework of the rib cage embedded in the thoracic wall.

The intercellular matrix of bone also plays an important physiological role in the storage of calcium, which can be withdrawn from the bone when the calcium level in the blood is lowered. The bone marrow contains tissue that plays an important role in the formation of blood elements.

■ Examine a thin section of **compact bone** (figure 2.11) prepared from a long bone, such as the humerus or femur of a human or other mammal, and identify: (1) the central **Haversian canal** through which passes many small blood vessels and nerves; (2) the **concentric layers** of bone (lamellae); (3) the **lacunae,** or spaces that house the bone cells or osteocytes; and (4) the numerous fine **canaliculi,** which serve to interconnect the lacunae and the Haversian canals.

Muscular Tissue

Muscular tissue is specialized for **contractility** and therefore has the capacity to perform mechanical work. Three different types of muscular tissue are distinguished: **smooth muscle, skeletal muscle,** and **cardiac** (heart) **muscle.**

Smooth Muscle

Smooth muscle consists of elongated, spindle-shaped cells with a single, central nucleus. This type of muscle is sometimes referred to as nonstriated muscle because it lacks the cross striations seen in both striated and cardiac muscle.

Smooth muscle forms the simplest type of muscle tissue and is generally found in parts of the body not under voluntary control where rapid movement or contraction is not essential, such as in the walls of the digestive tract, in the walls of blood vessels, and in the walls of the urinary bladder and the uterus.

■ Examine a slide of smooth muscle that has been teased apart to show the individual spindle-shaped cells or a section through some smooth muscle that shows the individual cells cut longitudinally. Locate the tapered smooth muscle cells and note the location of the nucleus. Identify the cell membrane and the fine longitudinal threads, the **myofibrils,** in the cytoplasm. Observe also the demonstration slides of smooth muscle from other types of material.

Draw several smooth muscle cells in figure 2.12 and label the **nucleus, cell membrane,** and **myofibrils.**

Fig. 2.12 *Student drawing of smooth muscle cells.*

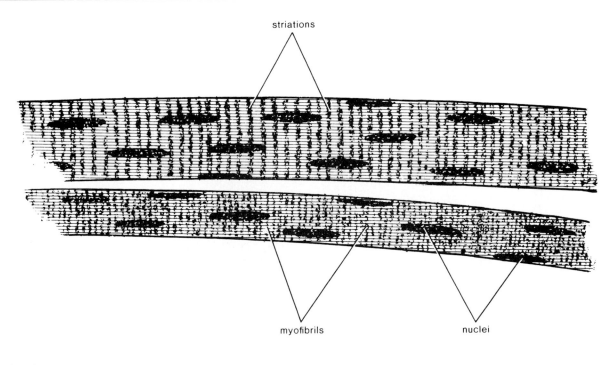

striations

myofibrils   nuclei

Fig. 2.13 *Striated muscle.*

Skeletal Muscle

The large muscles attached to various parts of the skeleton are composed of skeletal, or voluntary, muscle (see figures 2.13 and 2.14). This type of muscular tissue is made up of long cylindrical fibers containing many nuclei. What is unusual about the location of the nuclei in skeletal muscle as compared to most other kinds of cells? The multinucleate (syncytial) condition in skeletal muscle arises during embryonic development of the muscle as a result of the fusion of many mononucleate cells.

Note the conspicuous cross striations in these fibers and the outer limiting membrane, called the **sarcolemma.** Observe also the fine longitudinal **myofibrils** running lengthwise through the skeletal muscle fibers.

The individual myofibrils of striated muscle also have a very distinctive banded or striated structure. Because of their small size, individual myofibrils can be studied only with an electron microscope. Figure 2.15 is an electron micrograph showing parts of four adjacent myofibrils.

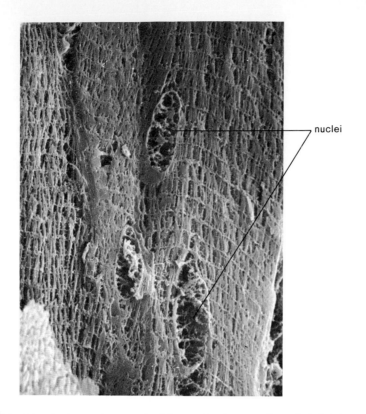

nuclei

Fig. 2.14 *Skeletal muscle. Scanning electron micrograph of freeze fracture section of skeletal muscle. Note muscle fibrils, cross striations, and ovoid spaces indicating location of nuclei near surface of muscle fibers. Magnification 1,850✕. (Photograph by Louis de Vos.)*

mitochondrion

Z-line

sarcomere

Z-line

Fig. 2.15 *Striated muscle fibrils. (Electron micrograph by Kenneth E. Muse.)*

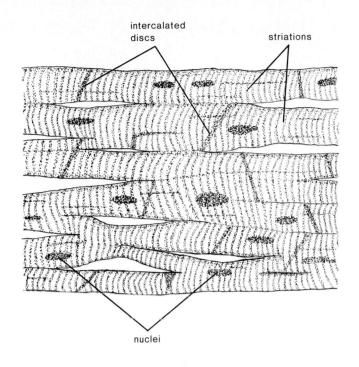

intercalated discs

striations

nuclei

Fig. 2.16 *Cardiac muscle.*

The functional unit of the myofibril is the **sarcomere,** which extends between two adjacent Z-lines. During muscle contraction, the sarcomeres shorten (the distance between Z-lines decreases). Recent investigations have shown that the contraction of muscle is due to interactions between two contractile proteins, **actin** and **myosin.** These two proteins make up a substantial portion of each myofibril.

The large **mitochondria** provide energy in the form of ATP (adenosine triphosphate) to power the contraction.

Cardiac Muscle

A third type of muscle tissue, **cardiac muscle,** is found in the walls of the heart of vertebrate animals. Cardiac muscle (figure 2.16) consists of striated muscle fibers, which branch and reunite (anastomose) with other fibers to form a continuous network of muscle fibers. Cardiac muscle contains cross striations like striated muscle, but the fibers of cardiac muscle are divided into cell-like units by many **intercalated discs,** which partially divide the fibers. This type of muscle, like smooth muscle, is involuntary; and throughout the life of the organism, cardiac muscle contracts and relaxes rhythmically and automatically. In fact, the heart of some of the lower vertebrates, such as a frog or a turtle, can be removed from the body and placed in a physiological salt solution where it will continue to beat for many hours, or even days.

■ Study the demonstration slides of human and other vertebrate cardiac muscle and observe: (1) the branching and anastomosing network of muscle fibers, (2) the cross striations, (3) the scattered nuclei, (4) the outer sarcolemma, and (5) the intercalated discs.

Nervous Tissue

The nervous system of vertebrate animals consists of the brain, spinal cord, and the nerves. Nervous tissue consists of highly specialized cells which carry impulses from one part of the body to another. A nerve cell, together with its branches or processes, which in some cases are many centimeters long, is called a **neuron.** A neuron consists of a **cell body** containing the **nucleus** and two or more elongated **nerve processes.** In a **motor neuron,** which transmits impulses from the central nervous system to the muscles or other effectors, there are several short **dendrites** and one long **axon.** The dendrites carry nerve impulses toward the cell body, and the axon carries nervous impulses away from the cell body. A **myelin sheath** surrounds the axon.

Figure 2.17 is an electron micrograph showing a cross section of a neuron in a cat. Observe the thick, many-layered myelin sheath surrounding the axon. Note also the many small tubules cut in cross section within the axon.

■ Examine also a slide with a stained smear preparation of the gray matter from the spinal cord of a cow. The details of the individual neurons are more readily

myelin sheath   axon

Fig. 2.17 *Neuron in ganglion of cat. (Electron micrograph by Kenneth E. Muse.)*

axon   dendrites

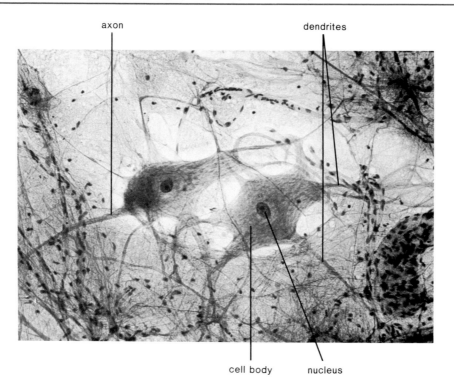

cell body   nucleus

Fig. 2.18 *Neurons from spinal cord smear from a cow. (Courtesy of Carolina Biological Supply Company, Burlington, N.C.)*

seen in such a preparation than in sections. Observe the large neurons (easily seen under low power). Compare with figure 2.18. Select a typical cell and note the **cell body, nucleus, nucleolus,** numerous **dendrites,** and the longer **axon.** Observe the numerous fine fibrils, the **neurofibrils,** within the cytoplasm of the cell body and extending into the processes.

Vascular Tissue

The blood of living vertebrates is a red liquid that is constantly in motion as it circulates through a closed system of tubes—the blood vessels. In humans, the blood comprises about 7 percent of the body weight. It consists of a

colorless liquid, the **plasma,** and several types of **blood cells** suspended within it. In permanent, stained preparations the plasma is not seen, and many of the cells, particularly the red corpuscles, may be distorted by the reagents used in the preparation of the slides.

Human Blood

Suspended in the plasma of human blood are the **red** and **white cells** and structures called **blood platelets.** Several types of human blood cells are shown in figure 2.19.

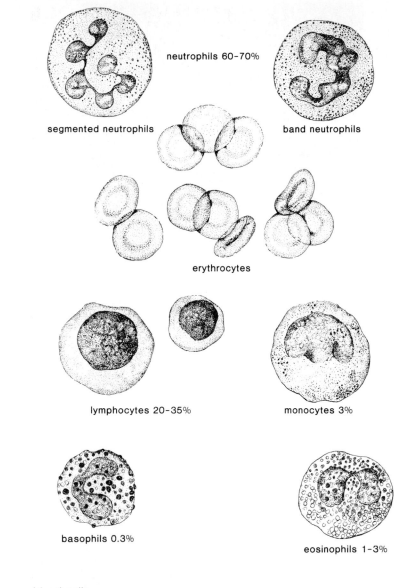

neutrophils 60-70%

segmented neutrophils

band neutrophils

erythrocytes

lymphocytes 20-35%

monocytes 3%

basophils 0.3%

eosinophils 1-3%

Fig. 2.19 *Types of human blood cells.*

■ Study the types of human blood cells in the prepared slides provided for you, or prepare your own blood smear slides by following the directions given at the end of this section.

1. **Erythrocytes.** Observe that human red blood cells, or erythrocytes, are small, circular, biconcave, and **lack nuclei.** Their chief function is to carry oxygen to the cells of the body. They contain large amounts of the iron-containing protein **hemoglobin,** which combines with oxygen in the lungs and releases it in the body tissues. The biconcave shape of human red blood cells is clearly shown in figure 2.20, which is a photograph taken with a scanning electron microscope. Figure 2.21 shows cross sections of two red blood cells within a capillary of a mouse in a transmission electron micrograph. Note the outer membrane and the uniform dark, granular cytoplasm of the erythrocytes. Observe the lack of nuclear material in both cells. The absence of a nucleus in mature red blood cells is characteristic of all mammals.

2. **Leucocytes.** Human blood contains several types of leucocytes or white blood cells. The types and relative frequencies of each type are shown in figure 2.19. Study your blood smear and identify as many types of leucocytes as you can find.

A cross section of a leucocyte of a mouse is shown in figure 2.22. Observe the large nucleus and the complex structure of its cytoplasm. Compare it with the relatively simple structure of the erythrocytes in figure 2.21. Why do you think that mammalian leucocytes would have a more complex structure than mammalian erythrocytes? How is their structure related to their function?

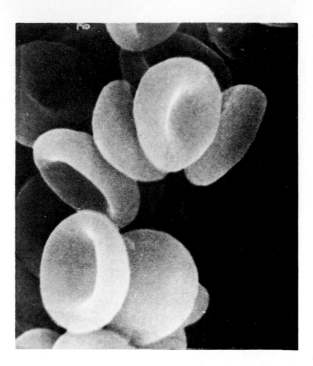

Fig. 2.20 *Human erythrocytes. (Scanning electron micrograph by Kenneth E. Muse.)*

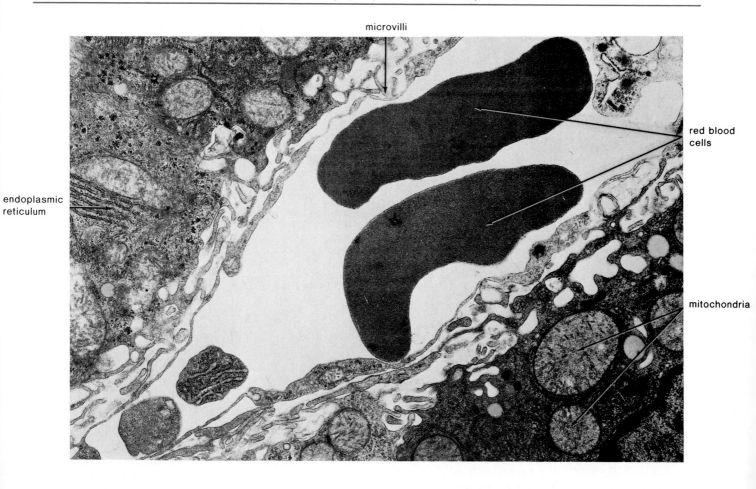

microvilli

red blood cells

endoplasmic reticulum

mitochondria

Fig. 2.21 *Erythrocytes in capillary of mouse liver. Magnification 13,750✕. (Electron micrograph by Kenneth E. Muse.)*

liver cell   lumen of capillary   leucocyte

nucleolus   nucleus   mitochondria

Fig. 2.22 *Leucocyte in capillary of a mouse. Magnification 21,000✕. (Electron micrograph by Kenneth E. Muse.)*

Figure 2.23 is a scanning electron micrograph of a **macrophage** from the human lung. Note the many irregular extensions from the surface of the cell. Macrophages are large phagocytic cells formed from monocytes which engulf and break down bacteria and cellular debris.

3. **Blood platelets.** Platelets are small, nonnucleated bits of cytoplasm that bud off from large cells (megakaryocytes) in the bone marrow. They are about one-third the size of erythrocytes and are not usually preserved in prepared microscope slides. Platelets function in **blood clotting** by temporarily plugging punctures that may occur in blood vessels and also by releasing substances that trigger later chemical reactions necessary for clotting.

Directions for Preparing and Staining a Blood Smear

1. Place a small drop of fresh blood about 1–2 cm from one end of a clean microscope slide.

2. Spread the drop of blood into a thin film covering most of the slide, using the end of a second clean microscope slide. Hold the second slide at about a 45° angle with the first slide and spread the drop of blood thinly over the surface of the first slide. Your laboratory instructor will demonstrate the technique for you.

3. Allow the blood smear to dry.

4. Lay the slide face up on a staining rack (or two glass rods) and cover the blood smear with a few drops of Wright's blood stain (a methylene blue-eosin mixture).

5. Stain for about 1–2 minutes.

6. Dilute the stain with about twice the volume of distilled water.

7. Drain the slide and rinse with distilled water until the thinner portions of the slide are pink.

8. Wipe the reverse side of the slide clean and let the slide air dry.

9. When dry, the slide may be examined under a compound microscope. A coverslip is not necessary, but one may be added for protection if desired. Consult your laboratory instructor for directions.

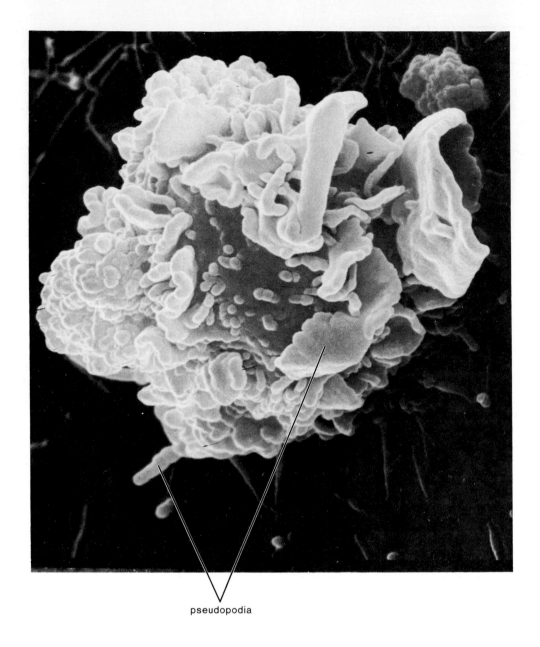

pseudopodia

Fig. 2.23 *Macrophage from human lung. Magnification 20,000✕. (Scanning electron micrograph by Kenneth E. Muse.)*

1. Ciliated epithelium (microscope slide).
2. Adipose (fat) tissue (microscope slide).
3. Fresh cartilage (from the sternum of a frog, ends of a long bone, etc.).
4. Cross sections (1–2 inches thick) of long bone to illustrate gross structure of bone.
5. Cardiac muscle (microscope slide).
6. Amphibian blood (microscope slide).

Key Terms

Bone hardened, mineralized type of connective tissue that provides support and protection in vertebrate animals.

Cardiac muscle type of muscle found in the heart wall of vertebrate animals. Consists of a branching and anastomosing network of striated, multinucleate muscle fibers.

Cartilage a firm and elastic type of connective tissue that provides support in vertebrate animals.

Cell membrane outer limiting membrane of an animal cell; composed of phospholipids, proteins, cholesterol, and carbohydrates.

Connective tissue type of animal tissue that binds, supports, and protects other body parts. Includes several kinds of tissue, such as loose connective (areolar) tissue, adipose (fat) tissue, cartilage, and bone.

Epithelial tissue type of animal tissue that lines the outer surface, and inner cavities and ducts of the body.

Erythrocytes red blood cells whose chief function is the transport of oxygen; contain large amounts of hemoglobin. May be nucleated (as in frogs and salamanders) or without a nucleus in the mature stage (as in humans).

Leucocytes white blood cells that are usually colorless in life and which perform many essential functions, including engulfment of foreign particles, production of antibodies, and wound healing. Several distinct types of leucocytes are usually present in the blood of an animal.

Muscular tissue type of tissue specialized for contraction; contains fibrils constructed of contractile proteins. Three types are distinguished—smooth muscle, skeletal muscle, and cardiac muscle.

Nervous tissue type of tissue specialized for the conduction of electrical impulses; important in the coordination of body activities. The basic functional unit is the neuron, or nerve cell.

Neuron the basic, functional unit of the nervous system of animals. The neuron is a single nerve cell, which consists of a cell body with a nucleus and two or more long extensions, or processes (axons and dendrites).

Nucleus the central organelle of an animal cell, which contains the genetic material (chromosomes) and which controls the metabolism of the cell.

Platelet a small, nonnucleated body in the blood of humans and other mammals. Formed by megakaryocytes in the bone marrow and serves mainly to plug leaks in blood vessels and to release chemical substances that initiate clotting.

Skeletal muscle type of muscle tissue found attached to parts of the skeleton. Consists of long, cylindrical fibers with a cross-banded or striated appearance in microscopic preparations; each fiber contains many nuclei. Skeletal muscle is also known as striated or voluntary muscle since many skeletal muscles are under voluntary control.

Smooth muscle type of muscle tissue found associated with internal organs in higher animals. Consists of spindle-shaped cells with a single central nucleus. Also known as involuntary muscle.

Vascular tissue made up of the blood cells and their suspending fluid, the plasma. Human blood consists of the red blood cells (erythrocytes), white blood cells (leucocytes), and platelets suspended in the plasma. Other animals may have additional types of blood cells.

3

Mitosis and Meiosis

Objectives

After completing the laboratory work in this chapter, you should be able to perform the following tasks:

1. Explain the basic differences between mitosis and meiosis.
2. Describe the cell cycle and explain the function of its principal stages.
3. Briefly describe the structure and function of chromosomes, centromeres, spindle fibers, aster rays, and centrioles.
4. Identify the principal mitotic stages of an animal cell in microscope slides or photographs.
5. Explain the chief events that occur in prophase, metaphase, anaphase, and telophase of mitosis.
6. Define random sample and explain its significance in biological research.
7. Describe a method for estimating the relative duration of the various stages of mitosis in a population of animal cells.
8. Explain the terms haploid and diploid and how they relate to the process of reproduction in animals.
9. Distinguish between a chromosome and a chromatid.
10. List the principal stages of meiosis and identify each stage in microscopic preparations or illustrations of animal cells.

Introduction

All animals depend on cell division for their growth and repair processes. Each cell has a precise set of genetic information built into its chromosomes. This information is essential for the proper functioning of each cell and for prescribing the characteristics of the next generation of cells. The division of the nucleus of the cell and the precise distribution of the duplicated chromosomes between the two new cells in this type of cell division is called **mitosis.**

Meiosis is a specialized type of cell division that usually occurs during the formation of the gametes or sex cells of multicellular animals. During meiosis, the normal diploid (2N) chromosome number of the somatic (body) cells is reduced by half to the typical (1N) chromosome number of the gametes. Meiosis is extremely important for the survival and evolution of animals because it provides for recombinations of genes during each generation. Thus, variations occur among the offspring of each generation and natural selection can operate to select the better adapted individuals.

Materials List
Prepared microscope slides
Whitefish blastula (mitosis) (or *Ascaris* embryo)
Ascaris ova (meiosis)
Audiovisual materials
Color transparencies of mitosis in whitefish blastula
Color transparencies of meiosis in *Ascaris* ova
Wall charts illustrating mitosis and meiosis

Mitosis

The division of nuclei by mitosis is exhibited by the somatic cells in most plants and animals. You should already have some idea of the complexity of cell division from your study of cell structure in an earlier laboratory session. During its life span, a cell passes through a regular sequence of physiological events called the **cell cycle**

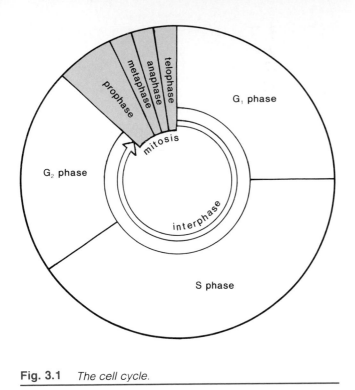

Fig. 3.1 *The cell cycle.*

(figure 3.1). This sequence includes several distinct stages, each characterized by certain metabolic activities of the cell. In actively dividing cells, the cycle may last only a few hours; in other cells, the cycle may last for days or weeks. At the completion of the cell cycle, a new generation of cells is produced by division of the parent cell.

The cell cycle is made up of four phases: G_1, S, G_2, and M. Together G_1, S, and G_2 constitute **interphase.** The **G_1 (gap) phase** is a period of active protein synthesis and the formation of new cell organelles, like mitochondria, Golgi bodies, ribosomes, and endoplasmic reticulum. It is a phase of rapid cell growth. The **S (synthesis) phase** is the time when DNA and other molecules making up the chromosomes, such as histones, are synthesized. Duplication of the DNA strands and of the chromosomes also occurs during S phase. In the **G_2 (gap) phase,** the cell synthesizes the proteins, actin and tubulin, plus the enzymes and other materials necessary for the mitotic spindle. **Mitosis (M phase)** follows the G_2 phase and involves the division of chromosomes and the formation of two new nuclei. Mitosis consists of four distinct stages, but actually takes up only about 5–10 percent of the complete cell cycle in most cells.

Abundant evidence now demonstrates that the duplication of the genetic material (DNA synthesis) in the nucleus occurs in the **nondividing cell prior to the initiation of mitosis.** The chromosomes are **already doubled** when they become visible during prophase, the first stage of mitosis. The term **mitosis** refers specifically to the process of nuclear division, the orderly distribution of the chromosomes between two daughter nuclei, starting with pro-

phase and continuing through metaphase, anaphase, and telophase. Technically, therefore, mitosis occurs **after** duplication has been completed.

Nuclear division usually occurs in close association with division of the cytoplasm (**cytokinesis**). The fact that these two processes do not always occur together, along with evidence from numerous experiments that have demonstrated that various chemical and physical treatments of dividing cells have different effects on mitosis and cytokinesis, clearly shows that different **chemical** and **physical** processes are involved in nuclear division (mitosis) and in cytoplasmic division (cytokinesis).

An important point to remember is that cell division is a **dynamic** series of events during which the cell undergoes dramatic and often rapid physiological and morphological changes. The so-called stages of mitosis merely represent a few morphologically identifiable points in this continuum.

The Mitotic Apparatus

During mitosis, a new structure, called the **mitotic apparatus,** is formed within the dividing cell (see figure 3.2). The mitotic apparatus plays an essential role in mitosis and provides the mechanism by which the duplicated chromosomes of the dividing cell are distributed between the daughter cells. A knowledge of the structure of the mitotic apparatus and its principal parts will help you to understand the process of mitosis and to learn how genetic continuity is maintained throughout many generations of cells. The mitotic apparatus is of vital importance to you; without it, a corn plant would not grow, a cut finger would not heal, a broken leg would not mend, and you would be unable to have children.

There are six main components of the mitotic apparatus: the **asters,** the **chromosomes,** the **centromeres,** the **centrioles,** the **centrosomes,** and the **spindle fibers.** Most dividing animal cells exhibit all six of these components. Most plant cells and certain invertebrate animal cells, however, lack centrioles and aster rays. Five of these six parts of the mitotic apparatus appear to play important roles in the process of mitosis, but the function of one of these structures, the asters, is still uncertain. The functions of these various components will be considered later in this exercise.

Several important events take place in a living cell during the early stages of mitosis. These events include:

1. The breakdown of the **nuclear membrane** and the mingling of the nuclear contents with the cytoplasm of the cell.
2. The condensation of the chromosomal material within the nucleus to form discrete, visible chromosomes.
3. The separation of the centrioles and their migration to opposite sides of the cell.
4. The formation of spindle fibers and aster rays.

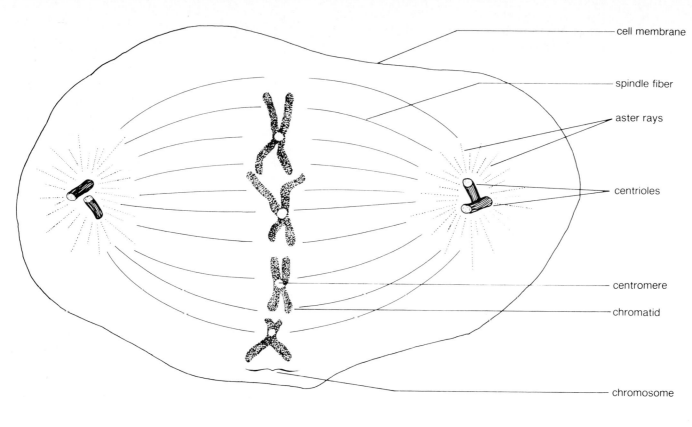

cell membrane

spindle fiber

aster rays

centrioles

centromere

chromatid

chromosome

Fig. 3.2 *Mitotic apparatus.*

The spindle fibers and aster rays are formed by the coalescence or condensation of relatively small protein molecules already existing within the cytoplasm of the cell. Thus, the formation of the components of the mitotic apparatus represents a recombination of molecules preexisting in the parent cell rather than the synthesis of new protein molecules. There is relatively little synthesis of new molecules during the process of mitosis; instead, most of the new molecules formed within the cell are synthesized during the interphase.

Chromosomes

Chromosomes are more or less elongate structures present throughout the life cycle of a cell. Their structure, however, appears quite different when observed at various times during the life cycle of a cell. During mitosis, the chromosomes can be seen as short, rodlike structures that are formed by condensation or contraction of very fine, threadlike structures present in the nucleus prior to mitosis. Thus, the appearance of chromosomes, like the appearance of the other components of the mitotic apparatus during the early portions of mitosis, results mainly from the reorganization of preexisting materials rather than from the actual synthesis of new materials.

Figure 3.3 is a photograph of the metaphase chromosomes in a male human cell with the normal 46 chromosomes (23 pairs). Observe the different sizes and shapes

Fig. 3.3 *Chromosomes from a male human lymphocyte in metaphase. (Photograph by Wendell Mackenzie.)*

of the chromosomes. Each species of animal (and plant) has a characteristic chromosome number. In humans, this number is 46; in the fruit fly *Drosophila*, it is 8; in the dog, it is 78; and in the cat, it is 60.

The chromosomes contain the hereditary units, or genes, which specify each of the characteristics of a cell and determine all of its capabilities. Thus, the chromosomes are responsible for the control of cell metabolism and enable cells to differentiate and to organize into specialized tissues, organs, and organ systems.

Centromeres

A centromere is a specific segment of a chromosome that serves to attach the chromosome to a spindle fiber by the formation of a protein complex called the kinetochore. During the latter portion of mitosis, the chromosome pairs move toward opposite **poles** of the mitotic spindle. Recent studies indicate that the chromosomes appear to be both pulled and pushed toward the poles by separate actions of the microtubular proteins of the spindle. Other studies have demonstrated that damaged chromosomes in cells injured by X rays or by chemical agents sometimes separate into two or more fragments. Those chromosomal fragments that lack centromeres do not become attached to the chromosome and do not move toward the poles. Such observations clearly demonstrate the important functional role played by the centromeres in mitosis.

Spindle Fibers and Asters

Special research techniques have been devised that allow scientists to isolate and to study the mitotic apparatus from living cells. Biochemical studies on such preparations of isolated mitotic apparatuses have demonstrated that the **spindle fibers** and **asters** are microtubules constructed mainly of two common contractile proteins, actin and tubulin. These proteins are involved in the movements of all eukaryotic cells. Some recent evidence indicates that the asters may influence the location of the cleavage plane in cytokinesis.

Centrioles

Centrioles are tiny structures usually found in pairs near the nucleus in nondividing animal cells with a well-formed nucleus. The centrioles migrate toward opposite sides of the cell during the early stages of mitosis and seem to form centers or foci for the spindle fibers and aster rays at each end of the mitotic apparatus. Surrounding each centriole is an amorphous area called the **centrosome,** which serves as the organizer for the formation of the mitotic spindle. Most plant cells have centrosomes that serve the same purpose but which lack centrioles.

Stages of Mitosis

An important point to remember is that mitosis involves a dynamic series of events during which the cell is undergoing dramatic, and often rapid, physiological and morphological changes. When studying prepared microscope slides of cells in mitosis, you are merely seeing cells that have been killed and stained at specific points in this continuous process. For convenience in describing the process, mitosis is usually divided into four main stages: **prophase, metaphase, anaphase,** and **telophase** (figure 3.4). The stage between successive mitoses is called **interphase.**

■ Many rapidly growing tissues provide good material for the study of mitotic cell division. Obtain a microscope slide prepared from some appropriate tissue such as the whitefish blastula, early embryos in the uterus of the roundworm *Ascaris,* or the skin of an amphibian tadpole. Study first one or more cells in **interphase.** Observe the structure of the nucleus and the arrangement of the chromatin material. Can you identify definite chromosomes? How many nucleoli do you find? Is the number of nucleoli the same in all interphase cells?

Observe other cells on the slide and select cells in each of the main stages of mitosis for further study. During your study, attempt to follow the sequence of stages as described below and try to visualize the changes that occur during the transition from one "stage" to the next. Draw on figure 3.5 each of the stages of mitosis listed below.

1. **Early prophase stage** during which the **chromatin** material shortens to form long, coiled, threadlike chromosomes.

2. **Middle prophase stage** with relatively thick chromosomes.

3. **Late prophase stage** in which the chromosomes are further shortened and thickened. Under high magnification, late prophase chromosomes can be seen to consist of two separate strands, the chromatids, joined by a single centromere.

4. **Metaphase stage** showing the chromosomes arranged in a disclike pattern on the equatorial plane and attached to fibers of the mitotic spindle.

5. **Anaphase stage** during which the centromeres divide and the two chromatids of each chromosome move apart to opposite poles. The anaphase stage is relatively brief.

interphase

early prophase

middle prophase

late prophase

metaphase

anaphase

early telophase

late telophase

daughter cells

Fig. 3.4 *Stages of mitosis.*

early prophase   middle prophase   late prophase

metaphase, polar view   metaphase, side view   anaphase

early telophase   middle telophase   late telophase

Fig. 3.5 *Student drawings of animal mitosis.*

6. **Early telophase stage** showing the full set of chromosomes at each end of the elongated cell and the beginning of a **cleavage furrow** around the middle of the cell.

7. **Middle telophase stage** in which the individual chromosomes start to uncoil and lengthen, and begin to appear less distinct.

8. **Late telophase stage** during which the nuclei in the two daughter cells reorganize. The chromosomes disappear, nuclear membranes and the nucleoli reappear, and the separation of the daughter cells is completed.

Timing in the Cell Cycle

The different stages in the cell cycle are not of equal duration; some stages are relatively long while others are very brief. The actual duration of mitosis and the relative duration of individual stages of mitosis vary widely from one cell type to another. Cells of a particular type also vary in their rates of division depending upon numerous physiological and environmental conditions. Nevertheless, you can obtain an estimate of the relative duration of the various stages in a population of cells and learn something about the kinetics of cell division by employing some relatively simple experimental techniques. You will need a whitefish blastula slide for this experiment.

Consider the whitefish embryo from which your mitosis slide was prepared as a **population of dividing cells** and see what can be determined about the kinetics of cell division, the relative duration of various stages of mitosis, and the relationship between mitosis and interphase in the life cycle of a cell. The whitefish embryo represents a population of relatively homogeneous cells; during the earliest stages of development, the embryo consists of few cells which divide more or less synchronously. As development continues and the number of cells in the embryo increases as a result of cell division, the degree of synchrony decreases and division becomes progressively more randomized.

Your whitefish embryo slide represents a slice through a fish embryo containing several thousand cells. It therefore represents a sample taken from a larger population of cells. Still, the number of cells on your slide is too great for you to count readily in the limited time available in the lab. Select a still smaller sample of cells from the embryo to yield some numbers that you can use to estimate some characteristics of the population of cells that made up the original whitefish embryo.

■ Take your whitefish embryo slide and select a random sample of fifty cells. How can you be sure that you obtain a **random sample** of cells? Record the number of cells in each of the four mitotic stages and those in interphase and carefully record the results of your count in your notebook.

Random sample

A *random sample* can be defined most simply as a sample from a population in which *every member* of the population has an *equal chance* to be included. Thus, a sample in which the individuals selected are determined by use of a table of random numbers (or the random number generator in a computer) would be a random sample. A sample in which the individuals selected are determined by taking every tenth individual would not necessarily be a random sample. Why?

Note: Count only cells in which the nucleus appears in the section and enough nuclear material (chromosomes, mitotic spindle, etc.) can be seen to allow accurate identification of the stage.

Since the cells in your slide were all killed (fixed) at approximately the same time, they represent a sample of the fish embryo cells stopped in action at a particular point in time. **Thus, the frequency of cells in the various stages is proportional to the relative duration of the stages.**

Construct a bar graph on figure 3.6 to illustrate the results of your count. Show the number of cells in your sample in each stage (prophase, metaphase, anaphase, telophase, and interphase). In which stage did you find the most cells? The fewest cells? Which stage, therefore, would you conclude is the longest in duration? The shortest? What are the relative lengths of the other stages as estimated from your sample?

Compare the data from your slide with that obtained by other students from their slides. How do their counts compare with yours? How much variation does there appear to be in the counts on different slides? What are some of the sources of **variation** that may account for the differences in your counts? Are your conclusions regarding the relative lengths of the various stages the same as those of other students in the class or do they differ? Why?

Now, combine the data from your count with that of the other students in your lab section to obtain estimates of the characteristics of another population of cells. The population represented by these pooled data consists of the cells in several (20–24) different whitefish embryos, and the fifty cells counted by each student represent a series of subsamples taken from that larger population of cells.

From these pooled data, construct another bar graph on figure 3.7 to show the number of cells in each stage. Estimate the relative lengths of the various stages in the new cell population from these pooled data and compare the results with those from your count of a single slide. How do your results compare with and how do your conclusions differ from those based on data from a single whitefish blastula slide? Explain.

Fig. 3.6 *Graph of the distribution of cells in various stages of division. Data from your slide.*

Fig. 3.7 *Graph of the distribution of cells in various stages of division. Pooled data from the whole class.*

Procedure

1. Select a random sample of 50 cells from your white-fish embryo slide.

2. Identify the stage of each cell in your random sample (prophase, metaphase, anaphase, telophase, or interphase).

3. Record the stage of each cell in your sample in the Notes and Sketches section at the end of this unit.

4. Construct a bar graph on figure 3.6 to illustrate the results of your count.

5. Study your bar graph and identify which stage of mitosis appears to be the longest (most frequent in your sample) and which stage is the shortest (fewest in your sample).

Meiosis

Meiosis is a special kind of nuclear division which ensures the constancy of chromosome number in the cells of succeeding generations of organisms. Sexually reproducing animals form male and female **gametes** at some point in their life history. **Fertilization** normally occurs at a later time in the life cycle and involves the fusion of male and female gamete nuclei. Thus, in order to maintain a constant number of chromosomes in successive generations (and to avoid doubling the chromosome number each time), some mechanism is necessary to provide a reduction (halving) of chromosome number between successive fertilizations. The process that results in the reduction in chromosome number is called **meiosis.**

The somatic cells of every species of animal have a definite and characteristic number of chromosomes. In man, this number is 46; in the fruit fly *Drosophila,* the number is eight; in the roundworm *Ascaris,* the number is four. This is referred to as the **diploid** (2N) number of chromosomes because the chromosomes are arranged in pairs. One member of each chromosome pair came from the father, and one chromosome came from the mother.

During the formation of gametes in animals, the number of chromosomes is reduced by half, and the resulting gametes have the **haploid** (1N) chromosome number. The subsequent fusion of the two haploid gametes (egg and sperm) during fertilization results in a return to the diploid chromosome number.

Meiosis generally consists of two successive nuclear divisions called the **first** and **second meiotic divisions** (figure 3.8). Meiosis differs in two important respects from ordinary mitosis.

1. The final number of chromosomes in a gamete resulting from meiosis is **only half** that of the parent cell, and each gamete receives only **one chromatid** from each homologous pair of chromosomes that was present in the original parent cell.

2. During the reduction in number, the chromosomes are **assorted at random** so that each gamete receives a chromatid from one or the other member of each homologous pair. This random assortment of genetic material during meiosis plays a very important role in heredity.

Homologous chromosomes are the paired chromosomes found in diploid cells that are very similar in size and shape, but differ both in origin (one comes from the father and one from the mother) and in genetic composition. (The father and mother usually contribute different sets of alleles to the offspring.)

Meiosis, like mitosis, is a dynamic process during which the cells are undergoing continuous changes. Nonetheless, a good understanding of the process can be achieved by describing it as a sequence of two nuclear divisions, each with four distinct stages and with an intervening interkinesis stage between the first and second meiotic divisions.

Principal Stages of Meiosis

First Meiotic Division (Reduction Division)

Prophase I. Many of the events in the first meiotic prophase are similar to prophase in mitosis. The chromosomes become visible as the very thin chromatin strands coil and condense. The nucleoli and the nuclear membrane disappear, and the mitotic spindle appears.

The key difference between the first meiotic prophase and prophase in mitosis is that, in meiosis, each pair of homologous chromosomes comes closely together in an intimate pairing process called **synapsis.** By the end of the first meiotic prophase, each homologous chromosome pair is seen as two double-stranded chromosomes closely held together. This structure, formed as a result of synapsis and consisting of two chromosomes with a total of four chromatin strands, is called a **tetrad.** Tetrads are found only during prophase I of meiosis. Note the six chromosomes at the end of prophase I as illustrated in figure 3.8. Each chromosome consists of two **distinct strands,** or chromatids, attached to **a single centromere.**

Metaphase I. At the onset of metaphase I, the synapsed chromosome pairs move together as a unit to the equatorial plane. Each chromosome pair is attached to a single spindle fiber by its **two adjacent centromeres.**

Anaphase I. The important difference between this stage and the corresponding stage in ordinary mitosis is that there is **no division of centromeres** in anaphase I of meiosis. The centromeres of the homologous chromosomes simply move apart, and the two double-stranded chromosomes of each pair migrate toward opposite poles. Thus, **half** of the chromosomes move to one pole, and **half** of the chromosomes move to the opposite pole. Note the three chromosomes migrating toward each pole in anaphase I, as shown in figure 3.8.

Telophase I. The chromosomes reach the poles of the mitotic spindle in each of the daughter cells, the spindle disappears, and new nuclear membranes appear around the reforming nuclei in the daughter cells. The chromosomes begin to elongate, gradually fade from view, and nucleoli reappear within the nuclei.

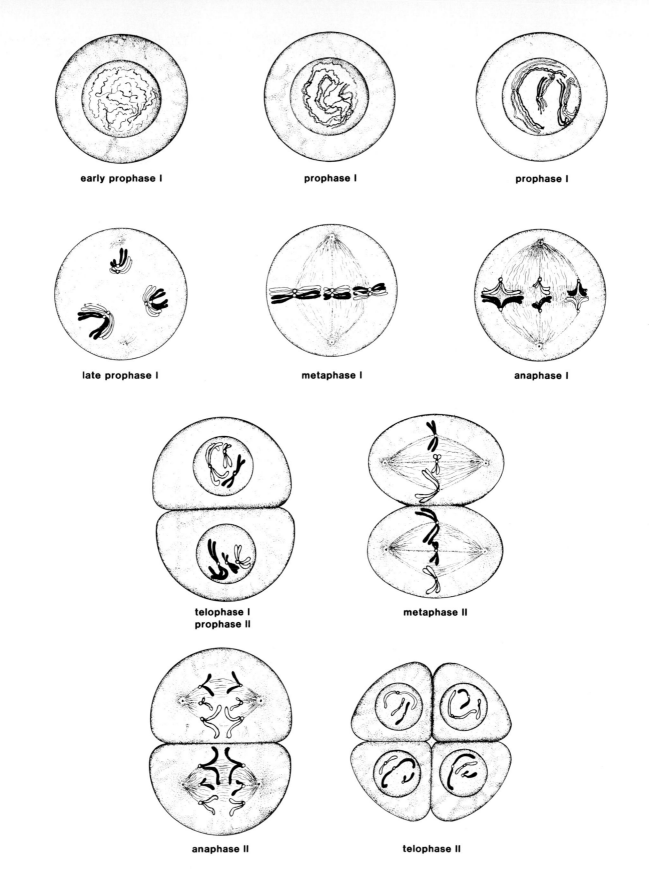

early prophase I   prophase I   prophase I

late prophase I   metaphase I   anaphase I

telophase I
prophase II   metaphase II

anaphase II   telophase II

Fig. 3.8 *Meiosis.*

Interkinesis. Between the two successive divisions in meiosis is a brief stage called **interkinesis**. This stage is generally similar to an interphase between mitotic divisions, but there is **no duplication of genetic material** (no DNA synthesis) in interkinesis.

Second Meiotic Division (Nonreduction Division)

Prophase II. The second meiotic division is essentially an ordinary mitotic division. There is no **synapsis** in prophase II; the double-stranded chromosomes reappear and move independently toward the equatorial plane.

Metaphase II. Each double-stranded chromosome attaches **separately** to a spindle fiber.

Anaphase II. The **centromeres divide** at the end of metaphase II, and during anaphase II the newly separated, single-stranded chromosomes move toward opposite poles.

Telophase II. The new chromosomes in the nuclei that appear in the daughter cells are **single-stranded** and contain only **half** the number of chromosomes as in prophase I.

The actual demonstration of meiosis to an inexperienced observer is difficult because of the small size of the chromosomes in most kinds of cells and the problems inherent in obtaining good microscopic preparations showing cells in clearly recognizable stages of meiosis. For these reasons, meiosis is often presented in introductory biology and zoology courses by means of a series of carefully selected demonstration slides. If possible, a time-lapse motion picture illustrating meiosis should also be seen to help you gain an understanding of the dynamic sequence of events in meiosis.

■ Study the demonstration materials on meiosis provided in the laboratory and **draw** selected stages of meiosis in the space provided in figure 3.9 as directed by your laboratory instructor.

Key Terms

Aster   includes all of the aster rays surrounding one pole of the mitotic apparatus in an animal cell. May serve to influence the location of the cleavage plane during cytokinesis. Absent in plant cells.

Aster ray   one of the fibrils, or rays, making up an aster.

Centriole   self-replicating tubular organelles usually found in pairs adjacent to the nucleus of an interphase cell surrounded by an amorphous centrosome. Also found centered in the asters of the mitotic apparatus of most animal cells.

Centromere   a narrow region of a chromosome that binds the two chromatids together prior to separation and which attaches the spindle fibers to the chromosome by the formation of a kinetochore during mitosis.

Chromatid   one strand of a duplicated chromosome.

Chromatin   genetic material in the interphase nucleus, representing the chromosomes in a long, thin, thread-like form.

Chromosome   filamentous structure that carries the genetic material of the cell (DNA). Chromosomes can be very long and uncoiled in interphase, shorter with double strands (chromatids) in prophase, and still shorter with a single strand in anaphase.

Cleavage furrow   indentation of the cell membrane around the equator of an animal cell at the beginning of cytokinesis.

Cytokinesis   division of the cytoplasm of a cell.

Diploid or 2N   cells containing both members of each homologous pair of chromosomes.

Equator   by analogy with the earth, a line that goes around the middle of the cell, equidistant from the poles.

Haploid or 1N   cells containing only one member of each homologous pair of chromosomes.

Homologous chromosomes   chromosomes with the same size and shape, and carrying the genetic material for the same characteristics. One member of each homologous pair comes from each parent.

Interkinesis   the period intervening between the first and second meiotic divisions. No chromosomes are duplicated and no DNA is synthesized during interkinesis.

Interphase   the stage between successive nuclear (mitotic) divisions, consisting of G_1, S, and G_2 phases. It is the stage during which the cells are metabolically active.

Meiosis   a special type of nuclear division in which the chromosome number is reduced from 2N to 1N by separating the members of the homologous pairs of chromosomes.

Mitosis   nuclear division resulting in two new nuclei with the same genetic complement as the original nucleus.

Mitotic apparatus   a special structure formed during mitosis consisting of the spindle fibers, asters, and centrioles.

Mitotic spindle   all of the spindle fibers collectively.

Nonreduction division   The second meiotic division. This division follows interkinesis and resembles a mitotic division. There is no reduction in chromosome number.

Nuclear membrane   (nuclear envelope)   the double membrane surrounding the nucleus in an interphase cell.

Nucleolus   a dense organelle within the interphase nucleus of eukaryotic cells; site of synthesis of ribosomal RNA and ribosome assembly.

Nucleus   a membrane-bound organelle containing the genetic material of a cell and which controls cell metabolism.

Poles   opposite ends of a cell where spindle fibers converge during mitosis and meiosis.

Reduction division   the first meiotic division, during which the number of chromosomes in a cell is reduced by half.

Spindle fiber   one of the microtubular filaments extending between the poles of the cell in mitosis and meiosis, made of contractile protein.

Synapsis   the pairing of homologous chromosomes in prophase I of meiosis, the centromeres touching each other.

Tetrad   a group of four chromatids of a pair of homologous chromosomes formed by a synapsis during prophase I of meiosis.

Fig. 3.9   *Student drawings of selected stages of meiosis.*

4

Development

Objectives

After completing the laboratory work in this chapter, you should be able to perform the following tasks:

1. Describe the structure of a spermatozoan of a frog or other representative animal, and identify its principal structures visible in a light microscope.

2. Briefly describe the major events in starfish or sea urchin development from fertilization to the gastrula stage and identify representative stages in a microscopic preparation.

3. Identify the principal structures in the blastula and gastrula stages of a starfish or sea urchin.

4. Discuss the organization of a frog egg and tell how it differs from a starfish egg.

5. Describe the major events in frog development from fertilization to the tadpole stage and identify representative stages from living or preserved specimens.

6. Discuss the organization of a chick egg and its adaptations for development on land.

7. Identify the four extraembryonic membranes surrounding a chick embryo and explain the function of each.

8. Identify the principal structures seen in whole mounts of 24- and 48-hour chick embryos.

Introduction

Animal development usually begins with the fertilization of an egg by a sperm. The nuclei of the egg and sperm fuse, and the male and female parents' genes share in determining the characteristics of the offspring. Compared to other biological processes, embryonic development is relatively slow. New cells, tissues, and organs make their appearance in the embryo over a period of hours, days, or weeks.

Embryonic development can be divided into five major phases: (1) **gametogenesis,** the formation of the haploid male and female gametes (sperm and eggs); (2) **fertilization,** activation of the egg and fusion of the sperm and egg nuclei to form the diploid zygote; (3) **cleavage,** the subdivision of the zygote into many cells

by mitosis; (4) **gastrulation,** the formation of germ layers; and (5) **organogenesis,** the initiation and differentiation of specific organs.

Materials List

Living specimens
    Frog embryos at various stages
    Frog sperm
Prepared microscope slides
    Starfish embryos
    Chick embryo, 24-hour, whole mount
    Chick embryo, 48-hour, whole mount
    Frog ovary with ova, cross section (Demonstration)
    Frog testis, cross section (Demonstration)
    Selected slides of early frog development (Demonstration)
Chemicals
    Amphibian Ringer's solution

In this exercise, we shall study examples of development from three different animals to illustrate different aspects of embryonic development and some variations in the development of different kinds of animals. The **starfish** illustrates the earliest stages of development, the **chick** serves to illustrate the later stages of development and adaptation for terrestrial life, and the **frog** is well suited for the study of development from fertilization to hatching.

Gametes

Gametes are the mature germ cells, **eggs** and **spermatozoa.** Both living sperm and stained microscope slides should be available for your study of male gametes. Observe the rapid movements of the living sperm in a wet mount.

■ On a stained microscope slide identify the anterior **head,** the narrower **middle piece,** and the long posterior **tail.** The sperm of different animal species vary considerably in size and shape, particularly in the shape of the head. Draw a sperm cell in figure 4.1 and label each of these parts.

Embryonic development consists of a complex series of processes by which a new organism arises from an egg. Following the fertilization of an egg with a sperm, the fertilized egg or **zygote** grows in size and increases in the complexity of its structure, function, and behavior. Different kinds of organisms exhibit many differences in the details of their development, but there are some important basic similarities in the development of all organisms.

We can identify four major component processes of development in organisms: **growth, morphogenesis, differentiation,** and **determination.**

Growth

Growth is the increase in mass of the organism through the addition of new cells and/or an increase in size of existing cells.

Morphogenesis

A second important process of development is **morphogenesis.** A living organism is not merely a bag of assorted parts heaped together in a random fashion. The parts of every living organism are arranged in a specific pattern and bear definite relationships to one another. Morphogenesis includes those processes through which the characteristic structures (both external and internal) of an organism become established. For example, the movements of cells and masses of cells in an embryo to form a wing or a limb are important processes of morphogenesis.

Differentiation

The progressive increase in the complexity of organization and specialization of individual cells and tissues is called **differentiation.** A fertilized egg of a frog, for example, forms many new cells and tissues during its development. This progressive increase in the number of biochemically and morphologically specialized cells and tissues is called differentiation.

Determination

An unfertilized egg has the potential to form a complete embryo with many kinds of cells, tissues, and organs. As development proceeds, and the egg becomes fertilized and divides to form new cells, individual cells and tissues become progressively restricted in the structures which they are able to form. This progressive limitation of the prospective fate of a cell or tissue is called determination.

Starfish Embryology

Starfish embryos are often used for introductory studies of development because they illustrate clearly the basic pattern of early development of multicellular animals (figure 4.2). Sea urchin embryos are also frequently used,

Chapter 4

Fig. 4.1 *Student drawing of sperm.*

Eggs, or ova, are much larger in size than sperm and contain varying amounts of stored food materials for the nourishment of the developing embryo. Since the eggs of the starfish, frog, and chick differ substantially in this regard, they will be described separately as we study the development of each animal.

Influence of Yolk

The pattern of development in animals is strongly influenced by the relative amount of **yolk** present in the egg. Based upon the amount and distribution of yolk, eggs can be classified into four main types: (1) **isolecithal eggs,** as found in starfish and humans, which have relatively little yolk and in which the yolk is uniformly distributed throughout the egg; (2) **mesolecithal eggs,** as found in the frogs, toads, and salamanders, which have a moderate amount of yolk and which have a concentration of yolk in the vegetal (lower) hemisphere; (3) **telolecithal eggs,** as found in birds and reptiles, which have a large amount of yolk and the cleaving portion of the embryo is restricted to a small disc at one end of the egg; and (4) **centrolecithal eggs,** as found in insects, which have much yolk and in which the actively developing portion of the embryo forms a thin layer around the outside of the large central yolk mass.

Among the animals chosen for this exercise, therefore, we have examples of three egg types:

Starfish egg—isolecithal
Frog egg—mesolecithal (although some textbooks classify frog eggs as "moderately telolecithal")
Chick egg—telolecithal

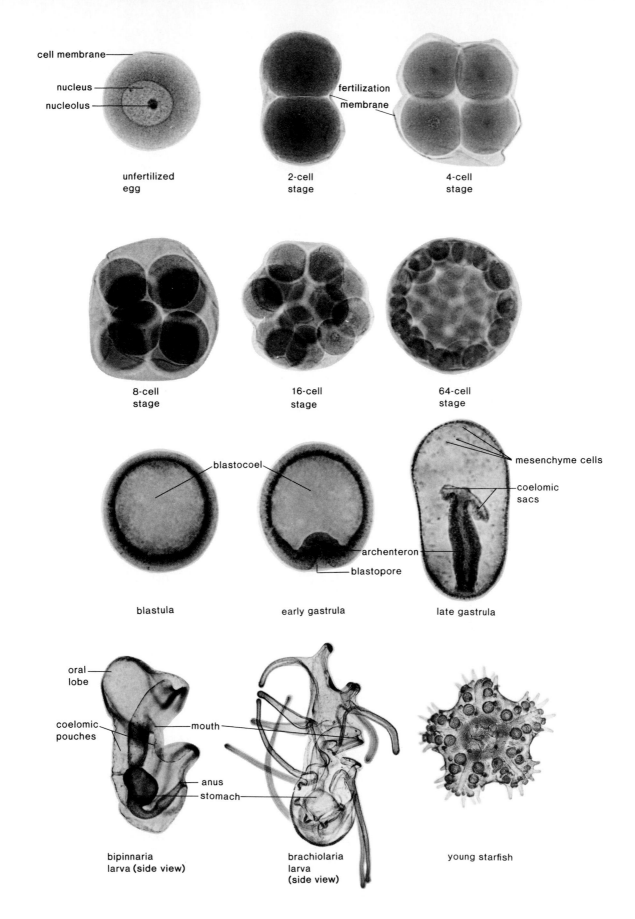

Fig. 4.2 *Starfish development. (Courtesy of Carolina Biological Supply Company, Burlington, N.C.)*

| (a) **early gastrula**
longitudinal section | (b) **gastrula**
longitudinal section | (c) **gastrula**
cross section |

Fig. 4.3 *Starfish gastrula.*

and their development is very similar to that of starfish in these early stages. Sea urchins also have the added advantage that living eggs and embryos are relatively easy to obtain for laboratory study. With living sea urchin embryos, it is possible to observe most of the early developmental events described later with an ordinary compound microscope.

■ Obtain a whole-mount microscope slide with stained starfish embryos at various stages of development. Examine the slide first under the low power of your microscope and select good representatives of the main stages described in the following section.

Color transparencies are excellent aids for the study of early embryology. They can be viewed easily in a small slide viewer or with the aid of a projector. Be sure to study the projection slides and any other demonstration materials available to assist in your understanding of starfish development.

Summary of Early Starfish Development

1. **Unfertilized egg.** Observe the large **nucleus** and the darkly staining **nucleolus.** A large food reserve, which provides energy for early development of the starfish embryo, is present as yolk in the cytoplasm. Surrounding the egg, or ovum, is a **cell membrane.**

   Prior to fertilization, the nuclear membrane breaks down, and the contents of the nucleus mingle with the cytoplasm. After penetration of the cell membrane by a spermatozoan, the haploid sperm nucleus migrates through the cytoplasm and fuses with the haploid egg nucleus. Also following fertilization, a new membrane is formed around the egg. This new membrane rises slightly above the surface of the egg and can often be observed in microscopic preparations. It is called the **fertilization membrane.**

2. **Two-cell stage.** Shortly after fertilization, the fertilized egg divides into two cells. This is the first of a number of rapid cell divisions that take place during the next several hours. This series of cell divisions is called **embryonic cleavage.**

3. **Four-cell stage.** Observe the fertilization membrane enclosing the four **blastomeres** (embryonic cells).

4. **Eight-cell stage.** What is the orientation of the cleavage plane that produced this stage relative to the last cleavage plane?

5. **Sixteen-cell stage.**

6. **Thirty-two-cell stage.** Note the rapidly diminishing size of the blastomeres. What has happened to the overall size of the embryo?

7. **Morula.** Solid mass of cells formed after many cleavages. No central cavity.

8. **Blastula.** Several more cleavage divisions occur, leading to the formation of a hollow ball of ciliated cells. This stage is the blastula, and after escaping from its enveloping membrane, the ciliated blastula swims about freely. The central cavity is the **blastocoel.**

9. **Early gastrula** (figure 4.3). A few hours later, the cells at one end of the blastula begin to push (or be pulled) into the blastocoel. This infolding of the embryo is called **invagination** and results in the formation of a two-layered structure. These two layers are **primary germ layers,** and similar layers are formed by virtually all multicellular animals. (Sponges are one of the few peculiar exceptions to this rule.) The layers are an outer **ectoderm** layer, which forms the skin and the nervous system of the

jelly coats

Fig. 4.4 *Unfertilized frog egg showing jelly coats. (Courtesy of Carolina Biological Supply Company, Burlington, N.C.)*

starfish, and an inner **endoderm** layer, which primarily forms the lining of the digestive tract of the adult starfish (and a few accessory organs of the digestive tract).

10. **Midgastrula** (figure 4.3b and c).   Invagination continues, and a hollow tube of endoderm extends into the blastocoel. The opening in the center of the endodermal tube is the **archenteron** (primitive gut) of the starfish embryo. The external opening of the archenteron is the **blastopore** (future anus of the starfish).

11. **Late gastrula.**   The endodermal tube later expands near its inner end and forms two lateral pouches. These lateral pouches become part of the internal body cavity or **coelom** of the starfish. The walls of these lateral pouches become differentiated and form the third primary germ layer, the **mesoderm.**

Frog Development

Ripe, unfertilized eggs of the grass frog, *Rana pipiens,* are relatively large (averaging about 1.75 mm in diameter) and have a moderate amount of yolk (mesolecithal type of egg). Remember that the frog is a semiaquatic

animal; its eggs must be deposited in water and development of the embryos and tadpole larvae can take place only in an aquatic medium.

■ Obtain a few unfertilized frog eggs in water and study them under your stereomicroscope (figure 4.6). Observe that the eggs are darkly pigmented. How is the pigment distributed on the surface of the eggs? Is the distribution of pigment uniform or nonuniform?

The center of the dark portion of the surface of the egg is called the **animal pole.** The center of the lighter surface (opposite the animal pole) is called the **vegetal pole.** The line connecting these two poles is the **animal-vegetal axis.** Shortly after fertilization the egg rotates so that the animal pole is uppermost. How does this differ from the orientation of the animal-vegetal axis of the unfertilized eggs in your sample?

Closely applied to the surface of the unfertilized egg is a transparent **vitelline membrane,** which lifts from the surface of the egg following fertilization and thereafter is known as the **fertilization membrane.** Surrounding the vitelline membrane are three **jelly coats** made up largely of albumen, a protein secreted by the oviduct during the passage of the egg from the ovary (see figure 4.4). The jelly coats serve to protect the egg from injury and infection.

Fig. 4.5 *Male and female frogs in mating position. (Courtesy of Carolina Biological Supply Company, Burlington, N.C.)*

Fertilization is external in the frog. During mating the male frog grasps the female with his forelimbs, and as the female discharges her ripe eggs into the water the male releases sperm over them (figure 4.5).

Shortly after fertilization, the vitelline membrane rises to form the fertilization membrane, and later a **gray crescent** area appears on one side of the egg along the margin of the heavily pigmented zone. The gray crescent is a more lightly pigmented zone of the egg cortex and appears on the side opposite the point of sperm entrance. The center of the gray crescent area marks the future **posterior end** of the embryo; thus, the future **anterior-posterior axis** of the frog embryo is determined at the time of sperm entry into the egg.

Development is fairly rapid following fertilization (see figures 4.6 and 4.7), and within a few hours at normal temperatures, the fertilized egg becomes divided into many cells, a hollow blastula is formed, and gastrulation is completed. Cleavage is **holoblastic** (all of the egg cytoplasm becomes divided) and **unequal** (the cells nearest the animal pole are smallest in size and those nearest the vegetal pole are largest in size). How is this related to the distribution of yolk in the frog egg?

A few hours after gastrulation, the first visible elements of the nervous system appear, first with the **neural plate** followed by the elevation of the **neural folds** that rise up on each side of the neural groove to form the **neural tube** within which the dorsal nerve cord begins to differentiate.

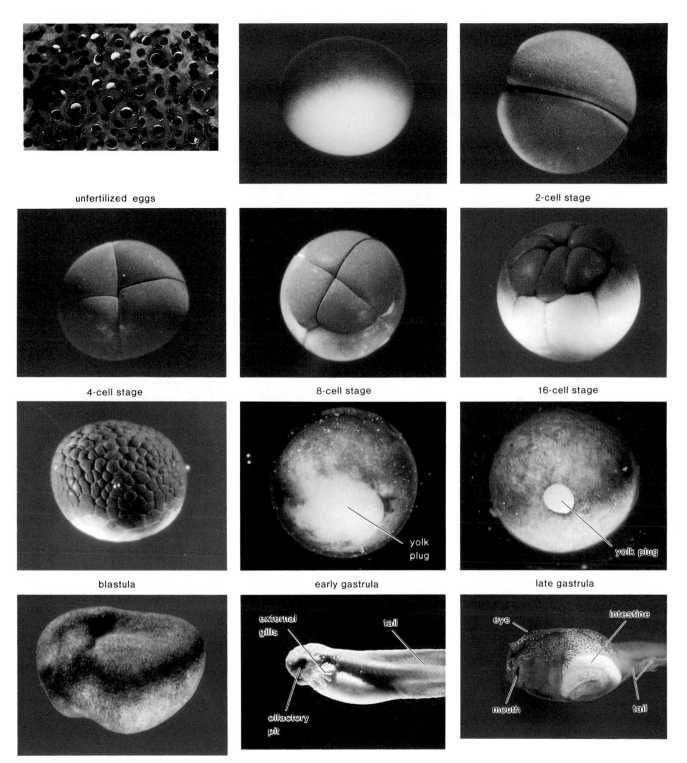

unfertilized eggs

2-cell stage

4-cell stage

8-cell stage

16-cell stage

blastula

early gastrula

yolk plug

late gastrula

yolk plug

neurula

early tadpole (lateral view)

external gills

tail

olfactory pit

late tadpole (ventral view)

eye

intestine

mouth

tail

Fig. 4.6 *Frog development. (Courtesy of Carolina Biological Supply Company, Burlington, N.C.)*

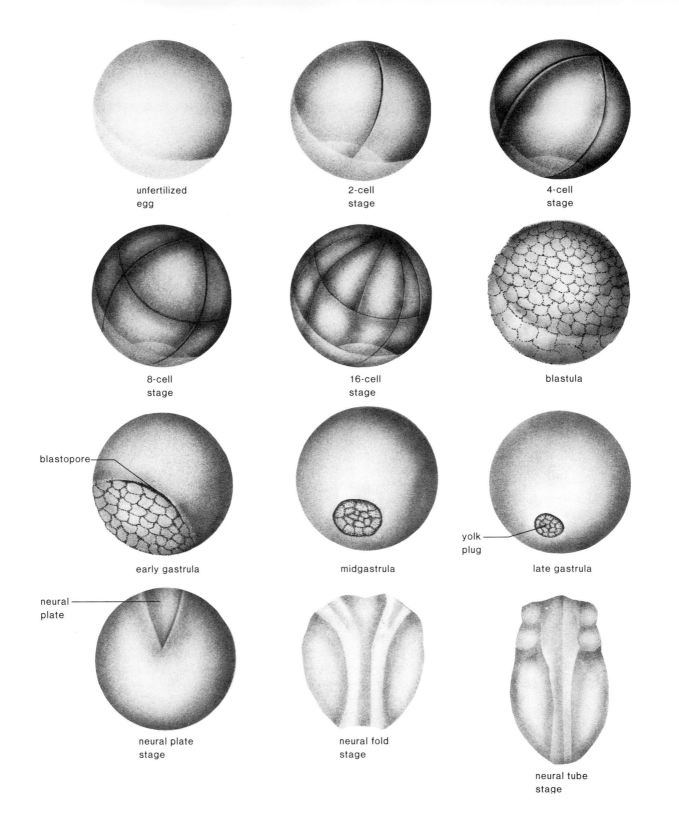

unfertilized
egg

2-cell
stage

4-cell
stage

8-cell
stage

16-cell
stage

blastula

blastopore

early gastrula

midgastrula

yolk
plug

late gastrula

neural
plate

neural plate
stage

neural fold
stage

neural tube
stage

Fig. 4.7 *Frog development.*

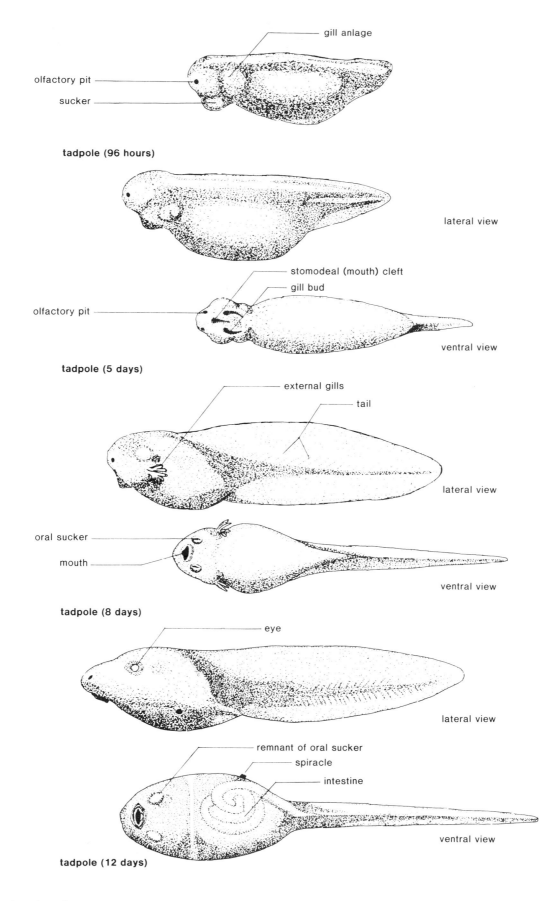

gill anlage

olfactory pit

sucker

tadpole (96 hours)

lateral view

stomodeal (mouth) cleft

gill bud

olfactory pit

ventral view

tadpole (5 days)

external gills

tail

lateral view

oral sucker

mouth

ventral view

tadpole (8 days)

eye

lateral view

remnant of oral sucker

spiracle

intestine

ventral view

tadpole (12 days)

Fig. 4.7 *(continued)*

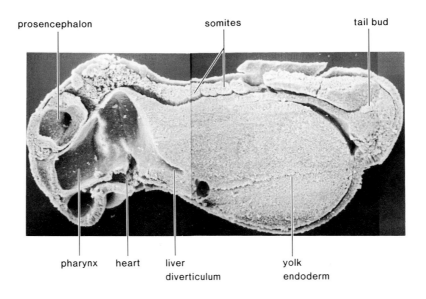

prosencephalon   somites   tail bud

pharynx   heart   liver   yolk
                    diverticulum   endoderm

Fig. 4.8 *Frog development. Sagittal section of the tailbud stage of a developing frog embryo. (Scanning electron micrograph of a freeze fracture preparation by Louis de Vos.)*

Table 4.1   Schedule of Development in *Rana pipiens* at 18° C (after Shumway and Rugh)

Stage	Hours at 18° C
Fertilization	0
Formation of gray crescent	1
Rotation (animal pole now uppermost)	1.5
First cleavage (2 cells)	3.5
Second cleavage (4 cells)	4.5
Third cleavage (8 cells)	5.5
Blastula	18
Early gastrula (dorsal lip stage)	26
Midgastrula	34
Late gastrula (yolk plug stage)	42
Neural plate	50
Neural folds	62
Ciliary movement	67
Neural tube formation	72
Tail bud stage	84
Muscular contractions	96
Heartbeat	5 days
Gill circulation, hatching (ruptures fertilization membrane)	6 days
Circulation in tail fin	8 days
Internal gills formed, opercular fold present	9 days
Operculum closed, covers internal gills	12 days
Metamorphosis into frog	3 months

Following the neural tube stage, the posterior portion of the embryo elongates to begin formation of the embryonic tail. This is the **tail bud stage** which occurs at about 84 hours postfertilization at 18° C (figure 4.8).

Table 4.1 illustrates the schedule of development in *Rana pipiens* at 18° C. Other common species of frogs, toads, and salamanders follow a generally similar pattern of development but differ in the times required to reach the various stages. Several representative stages of development in *Rana pipiens* are shown in figure 4.5.

■ Study the frog embryos provided in your laboratory and compare the various stages with figures 4.6 and 4.7. Living embryos are best for study if they are available, although plastic-embedded and preserved embryos can also be used. Supplement your study of the living or preserved whole embryos by observing the microscopic demonstrations provided to illustrate the internal structure of the embryos at different stages. When you have completed your study, you should be able to describe and explain the major events in the development of a frog from an unfertilized egg to an adult.

Demonstrations

1. Microscope slide showing developing eggs in the uterus.
2. Microscope slide showing frog testis with developing sperm.
3. Wet mount of living frog sperm in 10% amphibian Ringer's solution.
4. Microscopic slides with cross sections and/or sagittal sections of selected early developmental stages of the frog through hatching.
5. Procedures for inducing ovulation in the frog by injection of pituitary extract and artificial insemination of frog eggs in vitro.

Two-Day Embryo

Five-Day Embryo

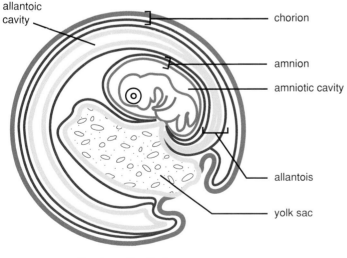

Fourteen-Day Embryo

Fig. 4.9 *Chick development, formation of extraembryonic membranes. Ectoderm is blue, endoderm is yellow, and mesoderm is red.*

Chick Development

The eggs and embryos of birds exhibit several major differences from those of amphibians. Birds have eggs adapted for development on land. This type of egg is called an **amniotic** (or cleidoic) **egg** because the embryo is enclosed in special extraembryonic membranes that protect and support the development of the embryo in a liquid environment. Other important adaptations of the eggs of birds (and reptiles) that allow them to develop on land are the inclusion of a large food supply and a hard outer shell. These features along with the extraembryonic membranes ensure an optimal environment during embryonic development.

Extraembryonic Membranes

Four extraembryonic membranes are formed by a developing chick embryo: **yolk sac, chorion, amnion,** and **allantois.** Each of these membranes forms in a specific way and each performs a distinctive role in protecting the embryo (figure 4.9).

The **yolk sac** forms as a pouchlike outgrowth from the developing gut. It grows around the yolk, releases enzymes to digest the yolk, and transports the digested yolk products through its blood vessels to the developing embryo.

The **chorion** and **amnion** are sheets of living tissue that grow out of and around the embryo. These sheets ultimately join above the embryo and enclose the embryo in a double sac. Both the chorion and the amnion consist of two tissue layers, ectoderm and mesoderm. After closure of the amnionic folds, the amnion becomes filled with a watery fluid. Thus, the amnion maintains an aqueous environment to protect the growing embryo from desiccation and serves as a physical protection for it.

The **allantois,** like the yolk sac, arises as a saclike outgrowth from the ventral surface of the gut. It has quite a different function, however. In birds, the allantois collects and stores metabolic waste products (largely crystals of uric acid). The allantois also grows and fuses with the chorion to form the **chorioallantoic membrane.** The chorioallantoic membrane is highly vascularized and facilitates the exchange of gases between the embryo and the external environment.

Whole Mount of 24-Hour Chick Embryo

■ Study first a whole mount of a 24-hour chick embryo. Consult figures 4.10 and 4.11, and study the whole-mount side under your stereomicroscope.

Caution: Whole mounts of embryonic stages are relatively thick and should never be viewed under high power on your compound microscope.

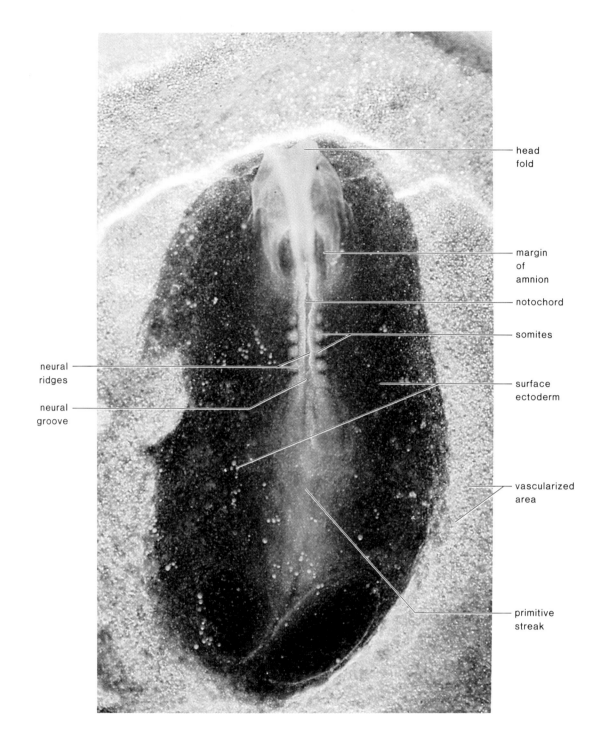

head
fold

margin
of
amnion

notochord

somites

neural
ridges

neural
groove

surface
ectoderm

vascularized
area

primitive
streak

Fig. 4.10 *Chick development: 24-hour embryo, dorsal view. (Courtesy of Carolina Biological Supply, Burlington, N.C.)*

Note the approximate size and shape of the embryo. Observe the transparent area immediately surrounding the embryo and an opaque outer area. The inner portion of this opaque area contains many blood vessels that bring nutrients to the developing embryo.

Locate the **headfold** of the chick at the anterior end of the embryo, lying free and slightly elevated above the underlying membrane. Within the head, observe that the brain at this stage consists of a **neural tube,** which is continued posteriorly with the open

Chapter 4

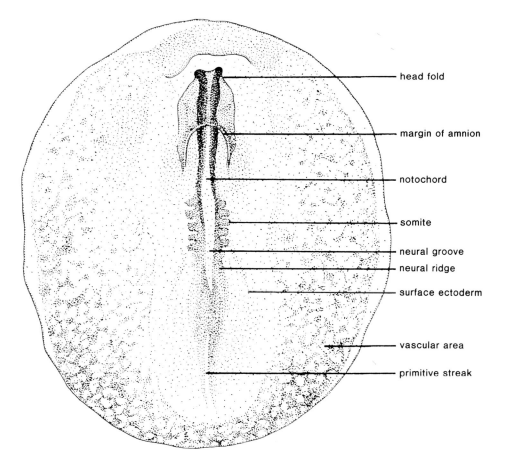

head fold

margin of amnion

notochord

somite

neural groove

neural ridge

surface ectoderm

vascular area

primitive streak

Fig. 4.11 *Chick development, whole mount of 24-hour embryo, dorsal view.*

neural plate consisting of a median **neural groove** and two lateral **neural folds.** Note also that the tissue of the neural folds is continuous laterally with the surface ectoderm, thus demonstrating the ectodermal origin of the nervous tissue.

Posterior to the neural plate is the primitive streak, the site of invagination of the surface ectoderm and the formation of the mesoderm layer. The primitive streak, thus, is functionally similar to or analogous to the blastopore of the frog and other amphibian embryos.

■ Find the **notochord,** which lies beneath the neural tube and the neural groove. Note that anteriorly, the notochord appears as a well-defined rod, but more posteriorly it appears as a wide band of less dense tissue.

Differentiation of the notochord, as of the neural structures, proceeds from anterior to posterior. Likewise, the **somites,** blocks of developing muscle tissue, differen-

tiate from anterior to posterior. New somites are formed by the coalescence and differentiation of mesoderm cells behind previously formed somites.

Whole Mount of 48-Hour Chick Embryo

■ Study a whole mount of a 48-hour chick embryo and observe some of its major features (figures 4.12 and 4.13). Note that the **head** and the **heart** are much more well-developed than at 24 hours. Observe that the **eyes** are clearly differentiated and that they exhibit a developing lens.

The brain is enlarged and is now divided into five divisions: the anterior **telencephalon** followed by the **diencephalon,** the **mesencephalon,** the **metencephalon,** and the **mylencephalon.** On the sides of the mylencephalon are a pair of **otic vesicles** that later form the ears.

The heart in a 48-hour embryo is a twisted tube with two well-established chambers: a **ventricle** and an **atrium.** The **sinus venosus** leading to the atrium is also beginning

metencephalon

ear

myelencephalon

left
vitelline
artery

mesencephalon

diencephalon

eye

prosencephalon

heart

posterior
margin
of amnion

somites

neural
tube

caudal
fold

Fig. 4.12 *Chick development: 48-hour embryo. (Courtesy of Carolina Biological Supply Company, Burlington, N.C.)*

Chapter 4

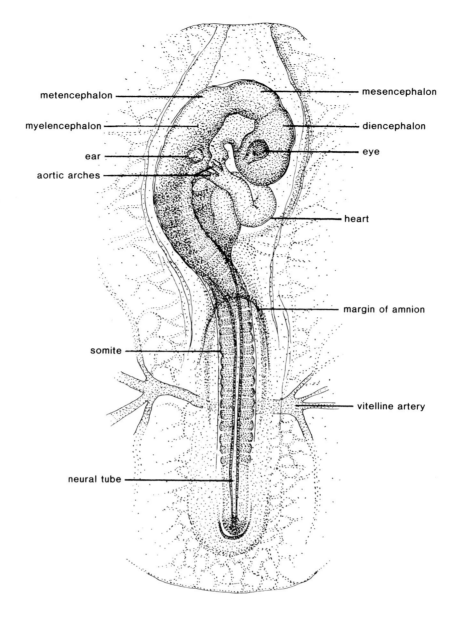

metencephalon

mesencephalon

myelencephalon

diencephalon

ear

eye

aortic arches

heart

margin of amnion

somite

vitelline artery

neural tube

Fig. 4.13 *Chick development, whole mount of 48-hour embryo.*

to develop. By 48 hours of development, the heart is actively pumping and circulation of the blood within the embryo and the surrounding vitelline area is well established. The heart connects with three pairs of large arteries, which represent the first three **aortic arches.** A large **dorsal aorta** extends posteriorly and connects laterally with two large **vitelline arteries.**

The head and the anterior part of the embryo are now covered by the **amnion,** which continues to grow posteriorly. The number of **somites** has increased to 24 pairs, which lie along each side of the neural tube. At the posterior end of the embryo, the caudal fold of the amnion has begun to grow forward to enclose the posterior portion of the embryo. Much later, the anterior and posterior portions of the amnion grow together and completely envelop the embryo.

Demonstrations

1. Living chick embryos after 33, 56, 72, and 96 hours of incubation (in chick Ringer's solution).
2. Microscope slides with selected cross sections of 33-hour chick embryos.
3. Microscope slides with whole mounts of 72- and 96-hour chick embryos.

Key Terms

Allantois one of the extraembryonic membranes found in the chick; forms as an outgrowth of the gut. Functions in gas exchange and for the storage of metabolic wastes.

Amnion an extraembryonic membrane found in the chick; consists of layers of ectoderm and mesoderm. Encloses the developing embryo and provides a fluid environment to protect the embryo.

Archenteron the primitive gut of an embryo; lined by endoderm tissue.

Blastopore the opening in the gastrula through which the ectodermal cells invaginate. Often becomes the mouth or the anus of the adult, depending on the type of animal.

Blastula hollow-ball stage in early embryonic development; follows the morula stage and precedes the gastrula stage.

Centrolecithal egg type of egg with abundant, centrally located yolk, such as an insect egg.

Chorion an extraembryonic membrane comprised of ectoderm and mesoderm layers; protects the developing embryo.

Cleavage the period of rapid cell division in an embryo following fertilization; leads to the formation of the morula and blastula stages.

Differentiation formation of a specialized tissue or cell type from a simpler tissue or cell type.

Ectoderm primary germ layer found on the exterior of an early embryo; forms the nervous system, integument, and certain other tissues in the adult.

Endoderm primary germ layer found lining the gut of an early embryo; forms the lining of the digestive tract, the liver, pancreas, thyroid, and certain other organs in the adult.

Extraembryonic membranes membranes that form external to the embryo of the chick and most other terrestrial vertebrate animals; serve to enclose and protect the developing embryo. Includes the chorion, amnion, allantois, and yolk sac in the chick.

Fertilization the fusion of the male and female nuclei, which initiates embryonic development. The sperm cell must first penetrate the cell membrane of the egg and migrate to the egg nucleus.

Gametes specialized sex cells, eggs and sperm, necessary for sexual reproduction.

Gametogenesis the formation of eggs and sperm; involves meiosis and cellular differentiation of the male and female sex cells to form gametes.

Gastrula developmental stage following the blastula. Invagination or inward movement of cells at this stage leads to differentiation of the three primary germ layers: endoderm, ectoderm, and mesoderm.

Gastrulation formation of the gastrula.

Growth increase in size or mass of an embryo or individual.

Invagination infolding of cells through the blastopore. Leads to the differentiation of endoderm and mesoderm.

Isolecithal egg type of egg with a small amount of yolk, as a starfish egg.

Mesoderm one of the primary germ layers; formed from invaginated ectoderm cells or as outpockets of the gut. Develops into the muscles, bone, connective tissues, circulatory system, and many other structures in the adult.

Mesolecithal egg type of egg with a moderate amount of yolk, as in the frog egg.

Morphogenesis the molding of a structure during embryonic development by cell and tissue movements.

Morula stage formed during the latter part of embryonic cleavage; consists of a solid ball of cells.

Notochord cartilaginous supporting rod parallel to the dorsal nerve cord of all chordates. May be replaced by a vertebral column in later stages of development.

Organogenesis the formation of a specific organ, such as the heart, during embryonic development.

Primary germ layer one of the three tissue layers differentiated early in development—ectoderm, mesoderm, and endoderm.

Somite a block of mesodermal tissue along the nerve cord in an embryo; differentiates into segmental muscles and other tissues in the adult.

Telolecithal egg type of egg with a large amount of yolk and the embryo restricted to one end, as a chick egg.

Yolk stored food reserves in the egg and embryo; rich in lipids and proteins.

Yolk sac one of the extraembryonic membranes of the chick and many other vertebrates; forms as an outgrowth of the gut and encloses the yolk.

Notes and Sketches

5

Protozoa

Objectives

After completing the laboratory work in this chapter, you should be able to perform the following tasks:

1. Identify the major structures in a specimen of *Euglena* and tell the function of each structure.
2. Describe the structure of a *Volvox* spheroid and of an individual *Volvox* cell.
3. Describe the reproductive processes and life cycle of *Volvox*.
4. Identify the principal structures in a specimen of *Amoeba proteus*.
5. Describe the techniques for preparing a wet mount and a hanging drop for microscopy and explain the uses of each.
6. Identify the main structures in a specimen of *Paramecium* and give the function of each structure.
7. Describe the functions of cilia in the feeding and locomotion of *Paramecium*.
8. Compare the locomotion of *Euglena, Amoeba,* and *Paramecium.*
9. Explain the life cycle of *Plasmodium* and identify the principal stages in microscope slides or photographs.
10. Explain the relationship of *Plasmodium* to the disease malaria.

Introduction

Protozoa are eukaryotic unicellular organisms, generally microscopic in size, that live as single individuals or in simple colonies. Within the unicellular body of a protozoon are many organelles that are analogous to the organs and organ systems of higher animals. Thus, protozoa exhibit a great deal of intracellular complexity. Most scientists believe that multicellular animals (metazoa) evolved from some group or groups of ancestral protozoa. Historically, there has been some disagreement among biologists over which group of protozoa may have been the real ancestors of the metazoa, but primitive flagellates are often cited as the most probable ancestral group.

For many years, biologists considered the protozoa to be the simplest group of animals and traditionally included them as the most primitive phylum in the animal kingdom. Recently, however, most biologists have come to recognize that the protozoa have more in common with other unicellular eukaryotic organisms than with multicellular animals. Therefore, protozoa are now usually placed in a separate kingdom, Kingdom Protista, with certain organisms formerly considered to be unicellular algae and fungi. Classification of the protozoa is further complicated by recent evidence suggesting that the differences among various groups are sufficiently important that they should be divided into separate phyla. A recently revised classification of the protozoa includes seven phyla.

NOTE: According to recent practice, the term protozoa is not capitalized since it no longer denotes a specific taxonomic unit.

Regardless of their taxonomic status, the protozoa are an important assemblage of organisms with many members that exhibit animal-like characteristics. For this reason and because of their possible phylogenetic significance as possible ancestors of multicellular animals, they should be included in any course in general zoology.

Some 70,000 species of protozoa have been described, including species widely distributed in many different kinds of moist or wet habitats; in fresh, marine, and brackish waters; in sewage; in moist soil; in or on the bodies of many species of animals; and in or on some plants.

In this chapter, we shall consider representatives of four common and well-differentiated groups of protozoa: the **Mastigophora**, the **Sarcodina**, the **Ciliophora** (formerly called Ciliata), and the **Apicomplexa.**

Classification

Phylum Sarcomastigophora

Protozoa with locomotion by means of flagella and/or pseudopodia; with a single type of nucleus.

Subphylum Mastigophora (Flagellata)

Locomotion by one or more flagella. Many members of this group are photosynthetic and exhibit other plantlike features. Examples: *Euglena, Volvox, Trypanosoma* (blood parasites of man and other animals), and *Gonyaulax* and *Gymnodinium* (dinoflagellates, often implicated in the red tides of coastal waters).

Subphylum Sarcodina

Locomotion by pseudopodia. Examples: *Amoeba proteus, Arcella* and *Difflugia* (testate amoebae), *Entamoeba histolytica* (a human parasite), *Globigerina* (a foraminiferan), and *Actinosphaerium* (a heliozoan).

Phylum Ciliophora (Ciliata)

Protozoa with cilia or ciliary organelles present in at least one stage of the life cycle; with two distinct types of nuclei (macronucleus and micronucleus). Examples: *Paramecium, Tetrahymena, Euplotes, Vorticella, Stentor, Blepharisma,* and *Trichodina.*

Phylum Apicomplexa

Protozoa typically lacking locomotory organelles (except for gametes in some groups); with a characteristic set of anterior organelles called the apical complex (visible only with the electron microscope); microspores present at some stage in the life cycle; all species parasitic. Formerly included in the Class Sporozoa.

Materials List

Living specimens
- *Amoeba proteus*
- *Euglena*
- *Volvox*
- *Paramecium caudatum*

Prepared microscope slides
- *Arcella* (Demonstration)
- *Difflugia* (Demonstration)
- *Entamoeba histolytica* (Demonstration)
- *Actinosphaerium* (Demonstration)
- *Globigerina* (Demonstration)
- *Peranema* (Demonstration)
- Symbiotic flagellates from termite or wood roach (Demonstration)
- Dinoflagellates (Demonstration)
- *Volvox*, cell walls (Demonstration)
- Flagellates illustrating Volvocine Series (Demonstration)
- *Paramecium*, pellicle (Demonstration)
- *Paramecium*, trichocysts (Demonstration)
- Representative ciliates (Demonstration)
- *Plasmodium* (Demonstration)
- *Eimeria* (Demonstration)

Chemicals
- Lugol's solution
- Protoslo (or methyl cellulose)

Miscellaneous supplies
- Congo red stained yeast cells

Audiovisual materials
- Chart of *Amoeba proteus*
- Chart of *Volvox* life cycle
- Chart of *Paramecium*
- Chart of *Plasmodium* life cycle
- Chart of *Eimeria* life cycle

A Solitary Flagellate: *Euglena*

Phylum Sarcomastigophora

Subphylum Mastigophora

Euglena (figure 5.1) is a common green flagellate often found in the greenish surface scum of standing or slowly moving water. *Euglena* is an enigmatic organism with a curious mixture of plant and animal characteristics, and, therefore, sometimes is considered to represent a borderline case between the plant and animal kingdoms. *Euglena* is smaller than *Amoeba* and *Paramecium* and, therefore, the details of its internal structure are more difficult to observe.

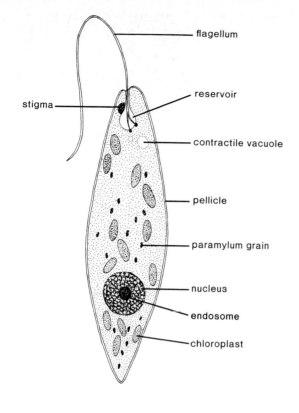

Fig. 5.1 Euglena.

Identify the following structures under high power on your compound microscope: (1) **pellicle,** the thick outer covering of the body; (2) **chloroplasts** with green chlorophyll; (3) **nucleus,** exhibiting a large central **endosome** in stained preparations; (4) **flagellar pocket;** (5) **contractile vacuole;** (6) a red **stigma,** or eye spot; (7) the long anterior **flagellum;** and (8) **paramylum grains,** a type of starch that represents stored food materials.

Euglena is quite sensitive to light, and changing the light intensity tends to cause the *Euglena* to move away. Bright light tends to make this protozoon remain stationary.

■ After you have completed your observations of the living specimen, add a drop of Lugol's solution (iodine and potassium iodide). This solution will kill the specimen and stain the flagellum to make it more readily visible.

As suggested by the presence of chloroplasts, the nutrition of *Euglena* is normally autotrophic; organic molecules (sugars) are synthesized from inorganic nutrients absorbed from the medium. Light from the sun provides the energy necessary for this process.

Biochemical tests have shown the paramylum granules to be a form of starch similar to that found in plants. Thus, both the presence of the chloroplasts and the storage of a plantlike form of starch indicate a close relationship of *Euglena* and its relatives to the plant kingdom.

■ Some species of *Euglena* are also able to survive, grow, and reproduce in the dark with no visible evidence of chloroplasts, chlorophyll, or stored food materials. How might such organisms obtain their food?

A Colonial (?) Flagellate: *Volvox*

Phylum Sarcomastigophora

Subphylum Mastigophora

Volvox (figure 5.2) is a common green alga that often occurs in great numbers in freshwater ponds and lakes. Although *Volvox* is photosynthetic and is sometimes considered to be an alga, it is often studied in zoology courses because it illustrates the organization of a simple colonial (or multicellular) organism well and because of its usefulness in illustrating one popular theory for the evolution of multicellular organisms from unicellular ancestors. It also demonstrates some basic similarities between plants and animals.

The spherical green *Volvox* are large enough to be seen swimming near the surface of a pond or in a laboratory culture even without a microscope.

■ Prepare a wet mount from a culture of living *Euglena* and observe the locomotion of an active specimen under your compound microscope. The active swimming movements result from the beating of the long flagellum, which pushes the organism through the water. A second, shorter flagellum is present within the flagellar pocket, but does not aid in the swimming movements. At certain times *Euglena* also exhibits another type of wormlike locomotion during which waves of contraction pass along the body in a characteristic fashion. This type of locomotion is peculiar to *Euglena* and related organisms, and is appropriately termed **euglenoid movement** or **metaboly.** It appears to result in part from the elasticity of the thick outer covering of the body, the pellicle.

After the wet mount begins to dry out, temporarily immobilizing some of your specimens, study the anatomy of a stationary *Euglena*. You will also find it useful to supplement your observations with the study of a prepared microscope slide.

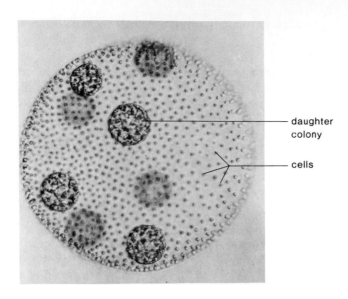

Fig. 5.2 Volvox, *asexual spheroid. (Courtesy of Carolina Biological Supply Company, Burlington, N.C.)*

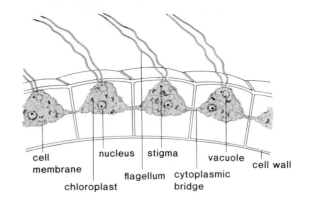

cell membrane   nucleus   stigma   vacuole   cell wall

chloroplast   flagellum   cytoplasmic bridge

Fig. 5.3 Volvox *cell structures, longitudinal section of cells at surface of spheroid.*

■ Obtain some living *Volvox* from a culture and prepare a wet mount. Be sure to add a few bits of broken coverslip or sand grains to protect the spherical *Volvox* bodies from being crushed by the weight of the coverslip. Observe the spheroid shape of *Volvox,* its swimming movements, and its prominent green color.

The spherical *Volvox* bodies are usually called colonies in textbooks, but recent studies have suggested that they are more similar to multicellular individuals. The coordination of cells and cellular function within the spheroid body is much greater than in most other colonial algae and protozoons; thus, recent workers use the term **spheroids** for the *Volvox* body.

The hollow *Volvox* spheroids average about 0.5mm in diameter and have many small green cells embedded in their outer walls (see figure 5.3). Each of the tiny body cells of *Volvox* contains a **nucleus,** a **contractile vacuole,** a green **chloroplast,** and two whiplike **flagella.** Some or all of the cells (depending on the species) may also have a red **stigma** (light-sensitive spot). The flagella project outward from the surface, and their beating keeps the spheroids in a constant spinning motion. Although the somatic cells of *Volvox* are very small and difficult to observe except with special microscopic preparations, the adjacent cells in some species of *Volvox* are connected by thin **cytoplasmic bridges** or strands. Other species of *Volvox,* however, lack these intercellular connections.

■ Add a drop of 0.1% methylene blue solution to a wet mount of *Volvox* and study under high power on your compound microscope. Can you observe any cytoplasmic bridges?

The green color of the spheroid individuals results from the presence of a chloroplast in each cell. What can you therefore infer about the nutrition of *Volvox?*

Within the spheroid you should be able to observe one or more large reproductive cells or **gonidia.** Reproduction in *Volvox* involves both sexual and asexual processes. In asexual development, embryos are formed from the gonidia. Locate several gonidia within the interior of an asexual parent individual (see figure 5.4).

During asexual development, the gonidia undergo a series of cell divisions remarkably like those seen in the embryonic development of many animal species. See if you can locate several different developmental stages of gonidia in the living specimens provided.

The beginning of sexual reproduction can be recognized when the gonidia form male and/or female spheroids. Most species of *Volvox* have separate male and female individuals (i.e., are **dioecious,** "two houses"), but some species produce both eggs and sperm in the same individual and are thus **monoecious** ("one house"). Figure 5.4 also shows both male and female individuals.

The living cultures available for study in the laboratory are usually all asexual. You should study the demonstration chart to learn about the life cycle of *Volvox.*

female
spheroid

asexual
spheroid

male
spheroid

Fig. 5.4   Volvox, *male, female, and asexual spheroids. (Photograph by Barbara Grimes.)*

■ Study the structure of *Volvox* using the living specimens provided in the laboratory. In a drop of water on a clean microscope slide, observe the swimming of *Volvox* spheroids first under low power, and then add a coverslip over the water drop and observe more details of the structure of the colony under higher magnification. Observe somatic cells, gonidia, eggs, sperm, and zygotes, using both the living specimens and the demonstration materials as necessary.

Evolution of Multicellularity

Volvox and several related green flagellates are often studied as models to illustrate one popular theory for the evolution of multicellular organisms. Most scientists believe that multicellular organisms arose from some unicellular form. The particular kind of unicellular organism is not known because this major evolutionary step took place more than 600 million years ago in the Precambrian Era. No well-preserved fossils have been found that actually document this transition from one to many cells, so biologists have searched among living plants and animals to seek **models** that might help their understanding of early evolution.

Volvox and several related green algae comprise the most popular model discussed by scientists. These related forms exhibit a graded series of solitary and colonial forms of increasing complexity and are called the **Volvocine Series** (figure 5.5). Among the important genera of algae comprising the series are: *Chlamydomonas, Gonium, Pandorina, Eudorina, Pleodorina,* and *Volvox.*

Other Mastigophora

Other important mastigophorans include the dinoflagellates, the symbiotic flagellates that inhabit the digestive tracts of termites and the wood roaches, some peculiar flagellates that may be related to sponges, and several important parasites of humans.

Dinoflagellates are found in both fresh and marine waters; many species form a characteristic outer covering, called a **test,** that is made of cellulose. Certain freshwater dinoflagellates may cause an unpleasant odor or taste in human water supplies. *Gonyaulax* and *Gymnodinium* are two marine dinoflagellates often associated with the red tides of coastal waters of North America, Europe, and Africa, that sometimes result in massive fish kills.

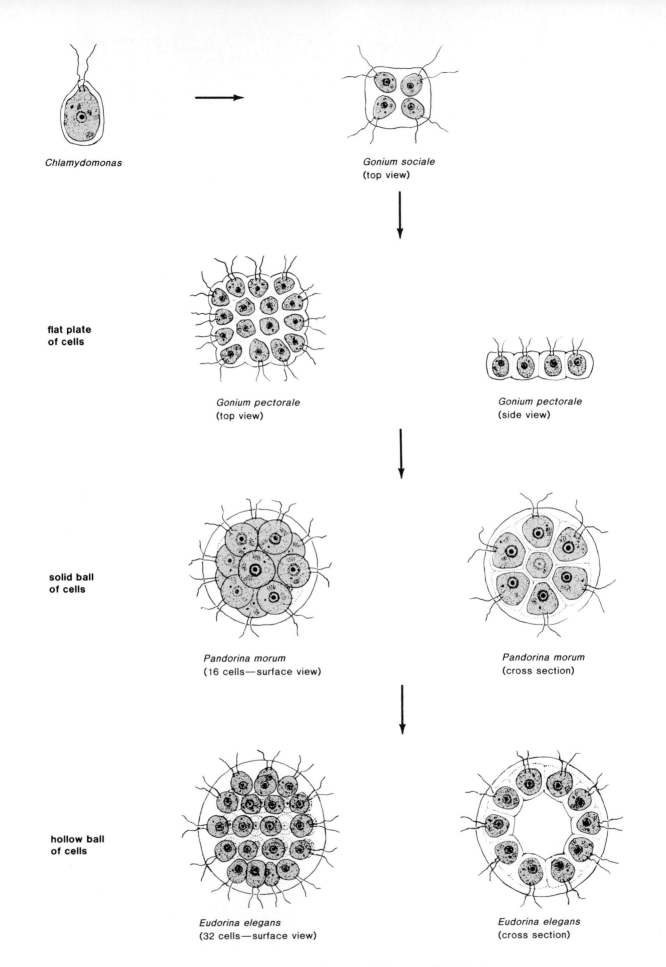

Fig. 5.5 Volvocine *flagellates illustrating possible evolution from unicellular to multicellular form.*

Fig. 5.6 *Symbiotic flagellates from termite gut. (Courtesy of Carolina Biological Supply Company, Burlington, N.C.)*

Some flagellates live as **symbionts** in the digestive tracts of wood roaches and termites (figure 5.6). Experiments have shown that the termites lack the digestive enzymes necessary to digest the cellulose in the wood they eat. The flagellates ingest splinters of wood and form food vacuoles around them. Later, digested products from the breakdown of the cellulose are released from the protozoa and provide nutrients for the termites. Termites from which the flagellates are experimentally removed soon die of starvation no matter how much wood they ingest. The flagellates benefit from the continuous supply of cellulose and from the suitable anaerobic environment of the host hindgut. Such a mutually beneficial symbiotic relationship is called **mutualism.**

Proterospongia is a colonial flagellate with species that closely resemble the flagellated collar cells, or choanocytes, characteristic of sponges (see chapter 6). Some biologists have suggested that the sponges may have evolved from some ancient protozoon similar to *Proterospongia.*

Still other mastigophorans are **parasites.** *Trypanosoma* and *Leishmania* are two important genera that include several serious human parasites. *Trypanosoma* has a thin, undulating membrane connecting its long, whiplike flagellum with its body (see figures 5.7 and 5.8). Several species of *Trypanosoma* cause sleeping sickness and other diseases in humans. *Leishmania* includes species that cause severe diseases in tropical areas of Africa, Asia, and South America.

Demonstrations

1. Large flagella in other Mastigophora, such as *Peranema.*
2. Microscope slide of trypanosomes.
3. Microscope slide of symbiotic flagellates from digestive tract of termite or wood roach.
4. Microscope slide of dinoflagellates.
5. Microscope slide showing cell walls in *Volvox.*
6. Chart illustrating the life cycle of *Volvox.*
7. Microscope slides with related volvocine flagellates, such as: *Gonium, Pandorina, Eudorina,* and *Pleodorina.*

An Amoeba: *Amoeba proteus*

Phylum Sarcomastigophora

Subphylum Sarcodina

Amoeba proteus (figure 5.9) is a protozoon found in ponds and streams. It often occurs on the undersides of plant leaves, and among diatoms and desmids. The transparent amoeba constantly changes shape by extending pseudopodia, footlike extensions of the cytoplasm, which serve for locomotion and in food capture. *Amoeba proteus* feeds on bacteria, small algae, and small protozoons.

In feeding, an advancing pseudopodium flows over one or more food organisms to trap the food in a water-filled cup. The opening of the food cup then narrows until the food is completely enclosed in a food vacuole.

white blood cell

parabasal body | red blood cell | trypanosome | undulating membrane

Fig. 5.7 Trypanosoma, *blood smear. (Courtesy of Carolina Biological Supply Company, Burlington, N.C.)*

flagellum

undulating membrane

Fig. 5.8 Trypanosoma. *Magnification 3,525×. (Scanning electron micrograph by Louis de Vos.)*

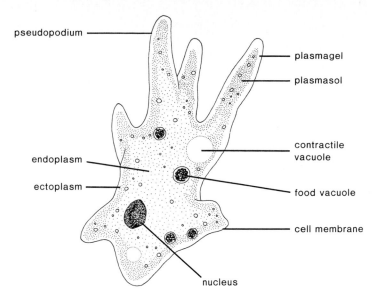

pseudopodium

plasmagel

plasmasol

contractile vacuole

endoplasm

ectoplasm

food vacuole

cell membrane

nucleus

Fig. 5.9 *Amoeba proteus.*

■ Prepare a wet mount to study living amoebae under your compound microscope.

Preparing a Wet Mount and a Hanging Drop. Obtain a clean microscope slide and add a drop of amoeba culture solution to the center of the slide. Take care to withdraw the drop of culture solution from the bottom of the culture dish or jar with a clean eyedropper or pipette. The amoebae are slightly heavier than the culture solution and are usually concentrated on the **bottom** of the vessel. Add a few bits of broken coverslip or grains of sand around the periphery of the drop to protect your specimens from being crushed, then carefully cover your preparation with a coverslip. This preparation is called a **wet mount.**

Another method of studying living Protozoa is to prepare a **hanging drop.** With this method you place a drop of the amoeba culture in a coverslip and invert the coverslip over the cavity of a depression slide. Both types of preparations can be observed for long periods of time if the outside edge of the coverslip is coated with petroleum jelly before it is placed in position on the microscope slide. Be sparing in your use of the petroleum jelly; avoid getting it into the culture droplet and on your microscope lens.

Study your preparation under the low power of the compound microscope to observe the general appearance of the amoeba. For best observation of a living amoeba, reduce the illumination to a minimum since living specimens are nearly transparent and almost invisible in bright light. Search your slide carefully for a specimen before discarding it or asking your instructor for a new preparation.

Locate an actively moving amoeba and note its constantly changing shape. The long, fingerlike projections are **pseudopodia** ("false feet"). Observe the lack of permanent orientation of the body of an amoeba; any portion may temporarily be anterior, posterior, right, or left.

With the aid of figure 5.9, identify and study the following structures found in the amoeba.

1. **Endoplasm**   the inner granular region, which forms the bulk of the cytoplasm.
2. **Ectoplasm**   the thin layer of clear cytoplasm which surrounds the endoplasm.
3. **Cell membrane**   the outer membrane surrounding the amoeba. Sometimes also called the plasmalemma.
4. **Plasmagel**   the stiff, jellylike, granular outer layer of colloidal endoplasm in the **gel state.**
5. **Plasmasol**   the central mass of colloidal endoplasm in a fluid, or **sol state.** Note the streaming movements within the plasmasol.
6. **Nucleus**   a transparent structure with no fixed position in the cell. It has the shape of a biconcave disc and often exhibits a folded or wrinkled appearance. Examine also the nucleus in a stained microscope slide of *Amoeba proteus.* Observe the darkly staining granular chromatin material within the nucleus.
7. **Contractile vacuole**   a clear vacuole found in the endoplasm which collects excess water from the surrounding cytoplasm and discharges it outside the

Fig. 5.10 Pelomyxa carolinensis. *A large multinucleate amoeba. (Courtesy of Carolina Biological Supply Company, Burlington, N.C.)*

body. Shortly after one contractile vacuole discharges its contents at the cell surface, a new contractile vacuole forms. Where are the new contractile vacuoles formed? Formerly, it was believed that the contractile vacuole also played an important role in the excretion of waste products from protein metabolism, but recent evidence has not supported this belief, and most specialists now agree that the contractile vacuole functions primarily in maintaining water balance in the cell (osmoregulation).

8. **Food vacuoles** vacuoles containing bits of ingested food and the digestive enzymes that act to break down these food materials into soluble materials that can be utilized by the amoeba. How are the food vacuoles formed? How are the undigested contents of a food vacuole disposed of after digestion has taken place?

Amoeboid Movement. *Amoeba* moves about by extending pseudopodia into which some of the innermost cell contents flow. Various kinds of amoebae form pseudopodia of different size and form. These pseudopodia are important in feeding, support, and locomotion. The mechanism of amoeboid movement has been studied by many scientists because of its intriguing nature and because similar movements occur in many other kinds of cells, including human leucocytes. Also, scientists now believe that amoeboid movement may be closely related to the phenomenon of cytoplasmic streaming, which occurs in virtually all kinds of living cells.

The movement of an amoeba is accomplished by the forward flow of the relatively liquid **plasmasol** from the center of the amoeba toward and into an expanding pseudopodium. Around the periphery of the pseudopodium, the plasmasol changes into a stiff **plasmagel.** Thus, the plasmasol moves the pseudopodium forward, and the plasmagel serves to fix it in position.

Recent biochemical and biophysical studies have demonstrated that the mechanism of amoeboid movement is similar to that in muscle contraction. **Contractile proteins** similar to the actin and myosin found in vertebrate muscles are present in the cytoplasm of an amoeba. Amoeboid movement results from folding, unfolding, polymerization, and depolymerization of these proteins.

■ Study the locomotion of an *Amoeba* on your microscope slide and also the pattern of its internal protoplasmic movements. In an active specimen, locate and carefully follow the movement of some granules in the plasmagel at the temporary posterior end. Observe how the plasmagel of the endoplasm changes into plasmasol, which flows forward and then changes into the gel state again, just back of the tip of the forming pseudopodium.

Reproduction. The reproduction of *Amoeba proteus* occurs only through the asexual process of binary fission. The nucleus and cytoplasm of a parent cell divide to form two daughter cells approximately equal in size. Thus, each of the daughter cells is genetically identical to the parent cell—excluding the rare occurrence of a mutation in one of the daughter cells.

Other Sarcodina

Many members of this group of Protozoa are more specialized than Amoeba. *Pelomyxa carolinensis* is a large multinucleate amoeba often studied in zoology classes (figure 5.10). Numerous species of amoebae live in shells

aperture

Fig. 5.11 *Test of a freshwater amoeba. (Scanning electron micrograph by F. W. Harrison.)*

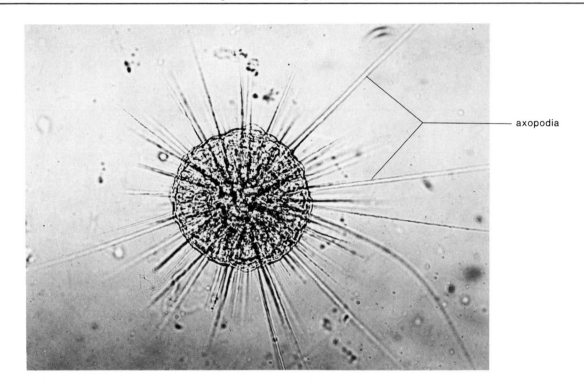

axopodia

Fig. 5.12 Actinosphaerium. *(Courtesy of Carolina Biological Supply Company, Burlington, N.C.)*

or tests which they secrete, or which they form from sand grains or other materials (figure 5.11). *Difflugia* and *Arcella* are two common testate amoebae found in freshwater ponds and streams. Other species of amoebae are parasites or symbionts in the digestive tracts of various animals. *Entamoeba histolytica,* an important parasite of humans, is the cause of amoebic dysentery, a disease often spread by drinking water or by eating raw vegetables contaminated by human wastes in parts of the world with poor sanitary facilities.

Some freshwater members of the Subphylum Sarcodina have many long, thin pseudopodia supported by axial rods of microtubules. *Actinosphaerium* (figure 5.12)

is a common freshwater example of a group called heliozoans because of the resemblance to the sun and its rays of sunlight.

Members of three marine classes of this subphylum, called **radiolarians,** form skeletons or tests of silicon and/or strontium compounds and exhibit many beautiful shapes (figure 5.13). The radiolarians are among the oldest known protozoa, and their tests are abundant in marine sediments in many parts of the world.

The **foraminiferans** (figure 5.14), representing another class of sarcodines, are an ancient and important group of marine sarcodines which form tests of calcium carbonate or other materials. The shells of foraminiferans

Fig. 5.13 *Radiolarian tests. (Courtesy of Carolina Biological Supply Company, Burlington, N.C.)*

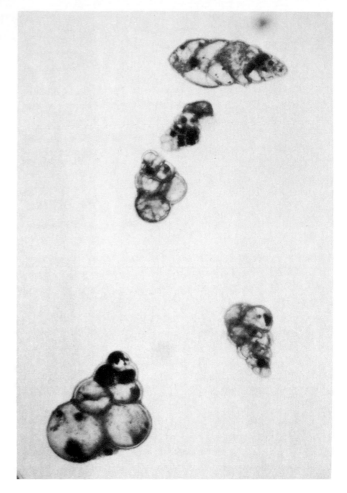

Fig. 5.14 *Foraminiferan tests. (Photograph by Barbara Grimes.)*

accumulate on the sea bottom and contribute to the formation of chalk and limestone. England's White Cliffs of Dover are made up largely of foraminiferan tests, as is much of the Bedford limestone found in Indiana and Illinois, and some of the limestone that was used to build the Egyptian pyramids. The distribution of certain species of foraminifera is also very important to petroleum geologists as an indication of ancient environmental conditions that may have been favorable for the formation of petroleum and thus as important clues to the location of petroleum pools.

■ Observe several of these other types of Sarcodina among the demonstrations.

Demonstrations
1. Models and charts of *Amoeba protcus*.
2. Culture of *Amoeba* under a stereomicroscope.
3. Microscopic slides with testate and parasitic representatives of the Subphylum Sarcodina, such as *Arcella, Difflugia, Entamoeba histolytica, Actinosphaerium* (a heliozoan), and *Globigerina* (a foraminiferan).

A Ciliate: *Paramecium caudatum*

Phylum Ciliophora (Ciliata)

Paramecium (figure 5.15) is a large, common, ciliated protozoon often found in water containing bacteria and decaying organic matter. There are several species of *Paramecium* that differ in various details of structure and that range in length from about 120–300 microns. The laboratory directions provided here are based upon *Paramecium caudatum,* a species frequently used for laboratory study and experimentation, but they will also apply, with minor modification (such as body size and number of micronuclei), to the study of other species of *Paramecium.*

■ Obtain a drop of *Paramecium* culture in a clean pipette and place it on a clean microscope slide with a similar-sized drop of methyl cellulose solution (or other similar agent) to slow movement. Methyl cellulose is a viscous material and serves mechanically

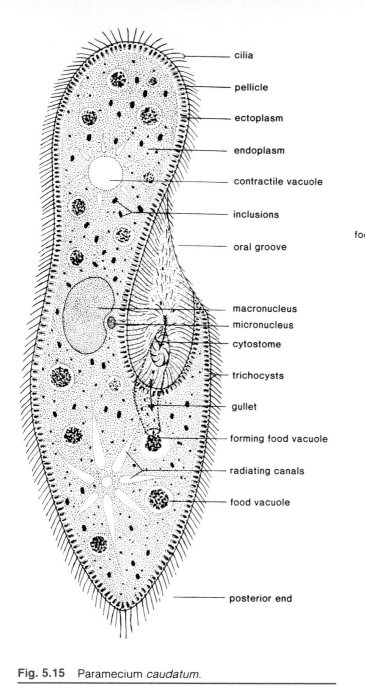

cilia

pellicle

ectoplasm

endoplasm

contractile vacuole

inclusions

oral groove

macronucleus
micronucleus
cytostome

trichocysts

gullet

forming food vacuole

radiating canals

food vacuole

posterior end

Fig. 5.15 Paramecium *caudatum.*

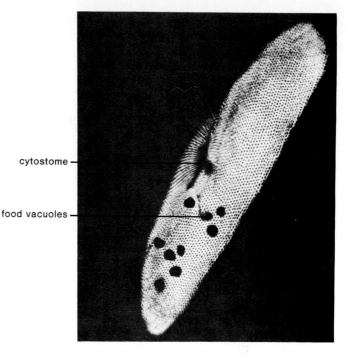

cytostome

food vacuoles

Fig. 5.16 Paramecium, *nigrosin stain to illustrate sculpturing of bilayered pellicle. (Photograph by Barbara Grimes.)*

to slow the swimming of the fast-moving *Paramecium.* Add a coverslip and observe your preparation under low power with your compound microscope. Note the form, color, and behavior of the animals in your preparation. Observe the slipper-shaped body with an **oral groove** beginning at the anterior end and running diagonally across the anterior portion of the animal. At the posterior end of this groove is the **cytostome,** or "cell mouth," through which food particles are passed as a result of the action of the specialized oral cilia lining the oral groove.

Paramecium is much more complex in its structure than *Amoeba.* Select a large, immobile, or slowly moving specimen, and with the aid of figure 5.15, identify and study the following structures.

1. **Cilia** the numerous cylindrical protoplasmic extensions that cover the surface of the *Paramecium* and that function in locomotion and in food gathering.

2. **Pellicle** the thick outer covering of the body through which the cilia project. The pellicle has a complex structure, but its details are difficult to observe without special techniques. Figure 5.16 shows some of the surface depressions in the bilayered pellicle.

3. **Trichocysts** tiny, rodlike structures embedded in the cortical (outer) cytoplasm beneath the pellicle. When properly stimulated, the trichocysts discharge their contents and form long threads. There is some evidence that the trichocysts may serve as a defense against predators, and they also serve to anchor the animal during feeding. In other types of ciliated protozoa, trichocysts have been found to have still other functions. Observe the microscopic demonstration of discharged trichocysts.

4. **Macronucleus** the large nucleus located near the center of the cell. Since it is transparent in a living animal, the structure of the macronucleus is best

studied in a stained preparation. Experiments have demonstrated that the macronucleus controls most metabolic functions of the cell.

5. **Micronucleus**   a smaller nucleus located close to and lying partly within a depression on the oral side of the macronucleus. The micronucleus is involved primarily in the reproductive and hereditary functions of the animal. This presence of two distinct types of nuclei is called **nuclear dimorphism** and is a condition found only in the Phylum Ciliophora. *Paramecium caudatum* has only a single micronucleus, but some other species of *Paramecium* have two or more micronuclei. As with the macronucleus, the structure of the micronucleus is best studied in a prepared microscope slide.

6. **Contractile vacuoles**   two clear, slowly pulsating vesicles located near each end of the body. Each contractile vacuole is surrounded by several **radiating canals** (not often seen in ordinary student preparations) which collect water from the surrounding cytoplasm. Observe the behavior of the contractile vacuoles. Are they fixed in position? Do they contract alternately or simultaneously? The function of the contractile vacuoles in *Paramecium* is the same as in *Amoeba,* the collection and discharge of excess water from the cell. Freshwater protozoa often have contractile vacuoles; marine protozoa generally lack them. How would you explain this difference?

7. **Cytostome** (cell mouth)   a permanent opening near the posterior end of the oral groove through which food is passed.

8. **Cytopharynx**   a short tube extending from the cytostome posteriorly and downward into the cytoplasm where food vacuoles are formed.

9. **Food vacuoles**   vacuoles located within the cytoplasm where they are carried by the streaming movements of the cytoplasm. Undigested materials are discharged through the **cytopyge,** or anal pore, located posterior to the oral groove.

Feeding

Paramecium is a filter-feeding organism and normally feeds on bacteria and yeast cells collected by a specialized food-collecting apparatus. An **oral groove** extends diagonally back along the body to a funnel-shaped **cytopharynx.** Food is swept along the oral groove by the action of specialized cilia lining the groove, is passed through the circular **cytostome** ("cell mouth") at the opening of the cytopharynx, and is passed through the cytopharynx into a newly forming **food vacuole.**

■ Prepare a wet mount with a drop of *Paramecium* culture to study the feeding process. Add a small amount of congo red stained yeast with the tip of a toothpick or clean dissecting needle. Try to pick up the smallest amount of yeast possible on the toothpick; too much yeast will cloud your preparation and obscure the *Paramecium.*

With this preparation, you can study the movement of the food particles, the formation of food vacuoles, and the subsequent movement of the food vacuoles within the cytoplasm. After the food vacuoles are formed, digestive enzymes are released into them and chemical digestion of the food particles begins. Note the color change in the vacuoles as the enzymes work. The color change is due to a change of pH in the vacuoles. Where do the digestive enzymes come from? Why don't they digest the other materials in the cell such as mitochondria and ribosomes? The diffusible products of digestion are released into the cytoplasm, and the undigestible remains are discharged at a specific site on the surface of the animal. This site is the **cytopyge,** or cell anus.

Cilia and Flagella

Most of the surface of *Paramecium* is covered by thin, hairlike projections called **cilia** (singular: cilium). **Cilia** are extensions of the cortical (outer) cytoplasm of the cell and play important roles in feeding and locomotion.

A great deal has been learned in recent years about the structure and function of cilia. These studies have revealed that cilia are closely related to the flagella (singular: flagellum) found on the surface of other kinds of protozoa. The structural differences between cilia and flagella are minor. When the projections are short and numerous, they are called cilia. When they are long and few, they are flagella. Cilia generally exhibit a relatively simple back and forth movement. The movements of flagella are often more complex and may involve a series of helical waves propagated along the flagellum.

Both cilia and flagella have a common basic structure. A cross section reveals an outer membrane enclosing a circle of **nine pairs of microtubules and two single microtubules** in the center of the cilium or flagellum. This basic pattern is found in all cilia and flagella, not only among the protozoa but also on the gills of molluscs, the ciliated epithelium lining the trachea of vertebrates, and the tail of spermatozoa.

Recent biochemical studies have also demonstrated that the movements of cilia and flagella involve **contractile proteins** similar to those found in striated muscle. This is another important illustration of the basic similarity of all living organisms.

Reproduction

Paramecium reproduces by a simple type of asexual reproduction in which the parent divides into two equal daughter cells. This type of asexual reproduction is termed **transverse fission** and is found in many kinds of protozoa.

Living specimens are occasionally seen in the process of fission, but the details of fission are best studied in a stained microscope slide. Obtain a prepared slide of *Paramecium* in fission and observe the nuclei. During fission, the micronucleus first divides by **mitosis,** and the macronucleus later divides by **amitosis.** No visible chromosomes are formed in the macronucleus; the macronucleus simply constricts, and the two portions separate. Macronuclear division is followed by cytoplasmic division

Micronucleus just divided into two, macronucleus elongated.

Two micronuclei, macronucleus dividing or constricted.

Macronucleus divided into two, cell body dividing.

Cell body almost constricted into two.

Fig. 5.17 *Student drawing of transverse fission in* Paramecium.

(cytokinesis). The process of fission may be completed rapidly, and under optimal conditions, *Paramecium* can reproduce asexually two or more times per day.

■ Draw the representative stages of fission in *Paramecium* as indicated on figure 5.17.

Unlike *Amoeba, Paramecium* can also reproduce sexually. The specialized type of sexual process exhibited by *Paramecium* is called conjugation (figure 5.18). During this process, two individuals come together, adhere by their oral surfaces, undergo a complex series of changes in both the macronuclei and the micronuclei, exchange a single pair of micronuclei (one from each cell), separate, and resume asexual reproduction. Following the exchange of micronuclei in each *Paramecium*, the newly introduced micronucleus fuses with another (nonmigrating) micronucleus. Thus, there is an exchange of hereditary material and a subsequent fusion of hereditary material from the two parents, a situation analogous to that of ordinary sexual reproduction studied earlier.

■ Examine a prepared slide or a demonstration of *Paramecium* in conjugation.

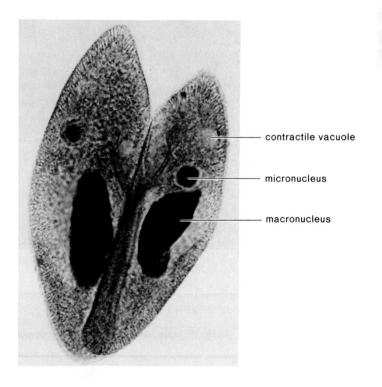

contractile vacuole

micronucleus

macronucleus

Fig. 5.18 Paramecium *in conjugation. (Courtesy of Carolina Biological Supply Company, Burlington, N.C.)*

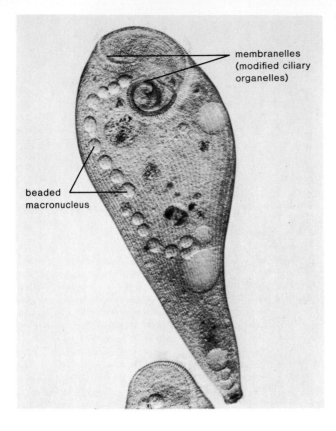

Fig. 5.19 Stentor. *(Courtesy of Carolina Biological Supply Company, Burlington, N.C.)*

Other Ciliata

The Phylum Ciliophora (Ciliata) is a large and diverse group of protozoa. One important form is *Stentor* (figure 5.19), a large, trumpet-shaped ciliate which is common in lakes, ponds, and streams, and which has been used in many experimental studies. *Stentor* has a spiral array of complex ciliary organelles leading to its cytostome and a beaded macronucleus. *Didinium* (figure 5.20) is a barrel-shaped predaceous ciliate with a voracious appetite. It feeds on other ciliates including *Paramecium*. A hungry *Didinium* can eat a *Paramecium* every two hours.

Spirostomum (figure 5.21) is a long, wormlike ciliate that has contractile fibrils that seem to function in a way similar to striated muscle fibrils. *Tetrahymena* (figure 5.22) is a small, ovoid ciliate that has been used in many experimental studies of biochemistry and genetics.

Vorticella (figure 5.23) is a sessile form with a long, contractile stalk which attaches to submerged stones, shells, plants, animals, and other objects. Some relatives of *Vorticella,* such as *Carchesium* and *Zoothamnion,* form stalked colonies.

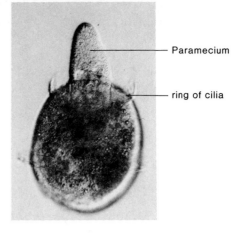

Fig. 5.20 Didinium, *a carnivorous ciliate ingesting a* Paramecium. *(Courtesy of Carolina Biological Supply Company, Burlington, N.C.)*

Podophyra is a suctorian, a specialized group of sessile ciliates that have **suctorial tentacles** in their mature stages and are predators of other ciliates. Cilia are found only on juvenile stages of the suctorians.

Demonstrations
1. Models and charts of *Paramecium*.
2. Stained slide showing pellicle of *Paramecium*.
3. Stained slide to show discharged trichocysts.
4. Stained slides with representative members of the Class Ciliata, such as *Stentor, Euplotes, Tetrahymena, Vorticella, Didinium, Blepharisma, Trichodina,* and *Podophyra*.

Phylum Apicomplexa

Members of this group were formerly included among the Sporozoa, but recent investigations have indicated that they should be considered a separate phylum. All members of this phylum are parasitic on other organisms.

Many species in this group parasitize invertebrate animals such as earthworms, crabs, and oysters, but the most important species are parasites of vertebrates, including humans. Among the most important sporozoans are *Plasmodium* and *Eimeria*. Several species of *Plasmodium* cause various forms of **malaria** in humans and other animals.

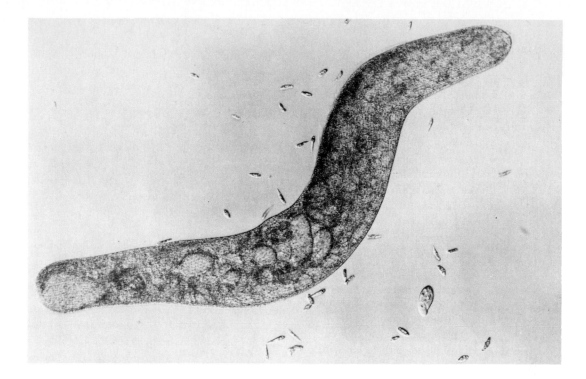

Fig. 5.21 Spirostomum, *a ciliate with contractile fibers and a flexible pellicle that exhibits a peculiar, wormlike locomotion. (Courtesy of Carolina Biological Supply Company, Burlington, N.C.)*

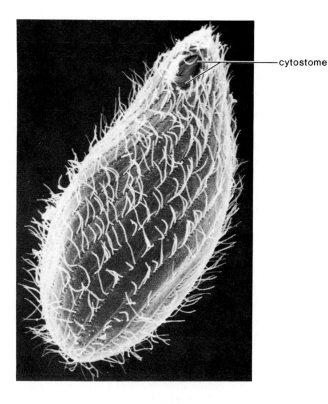

cytostome

Fig. 5.22 Tetrahymena. *(Scanning electron micrograph by Jolanta Nunley.)*

Eimeria is a genus of sporozoan parasites that causes **coccidiosis** in birds, rabbits, and other animals. This disease has great economic impact on the poultry industry. *Eimeria,* like *Plasmodium,* has a complex life cycle that includes both sexual and asexual forms. Study the demonstrations illustrating the life history and importance of *Eimeria* and coccidiosis.

The Malaria Parasite: *Plasmodium*

More than 50 species of *Plasmodium* have been described. All are parasites of vertebrate animals, including amphibians, reptiles, birds, and mammals. Four species cause human malaria, *P. vivax, P. ovale, P. malariae,* and *P. falciparum.* The life cycles of these species are all similar.

Malaria, one of the most serious and debilitating of human diseases, has had an important role in history from the fall of the Roman Empire to the war in Vietnam. Although modern medicine has made some progress in eliminating malaria, the disease is yet to be conquered. It is most prevalent in tropical and semitropical areas, and costs millions of lives and trillions of dollars annually.

Fig. 5.23 Vorticella. *A sessile ciliate with a long contractile stalk for attachment and a ring of cilia surrounding the mouth. (Courtesy of Carolina Biological Supply Company, Burlington, N.C.)*

Life Cycle of Plasmodium

The life cycle of *Plasmodium* (see figure 5.24) is complex like those of other apicomplexa and includes several generations with both sexual and asexual reproduction. The life cycle can best be understood by starting with the zygote in the gut of a mosquito, one of the two hosts necessary for the completion of the life cycle.

The zygote becomes motile and passes through the lining and wall of the stomach or midgut of the mosquito and is now called an ookinete. The ookinete then rounds up and encysts on the outside of the gut wall and is called an oocyst. After several days of growth, the oocyst divides internally to form several hundred sporozoites. The sporozoites escape by rupturing the external wall of the oocyst and migrate through the hemocoel to the salivary glands of the mosquito.

When a mosquito bites a human, the sporozoites and the mosquito's salivary secretions are injected into this host. Sporozoites that find their way into the human bloodstream are eventually carried to the liver, where the sporozoites enter host cells. Inside the host liver cells, the sporozoites transform into amoeboid multinuclear schizonts and feed upon the contents of the host liver cells. The schizonts reproduce asexually to form many merozoites (the next stage in the life cycle), escape from the liver cells, and, in some cases, invade other liver cells to repeat the process.

Merozoites escaping into the bloodstream penetrate erythrocytes to initiate the erythrocytic phase of the life cycle. Parasitic stages in the liver prior to entry into the erythrocytes are termed the exoerythrocytic phase.

The developing *Plasmodium* inside the erythrocytes exhibit a characteristic morphology, as seen in Giemsa-stained microscope preparations, and are recognized by their red nucleus and blue ring-shaped cytoplasm (see figure 5.25). This characteristic morphology is very helpful in the laboratory diagnosis of malaria. The merozoites in the erythrocytes undergo further asexual reproduction within the erythrocytes. Later, the merozoites rupture the wall of the erythrocytes, escape into the blood, and enter most erythrocytes. This multiplication process may be repeated several times, so that an enormous number of parasites are produced within the host. The rupture of the erythrocytes by the merozoites also releases accumulated toxic wastes from the parasites and causes the symptomatic chills and fever commonly associated with human malaria.

Some of the merozoites develop into sexual forms instead of repeating the asexual merogony. These sexual forms develop within the erythrocytes and become microgametocytes (male) and macrogametocytes (female) (figure 5.26). These stages are the progenitors of the male and female gametes, and represent the start of the sexual portion of the life cycle.

If a mosquito bites an infected host and ingests infected erythrocytes, the gametocytes pass into the mosquito's stomach, where they mature into microgametes and

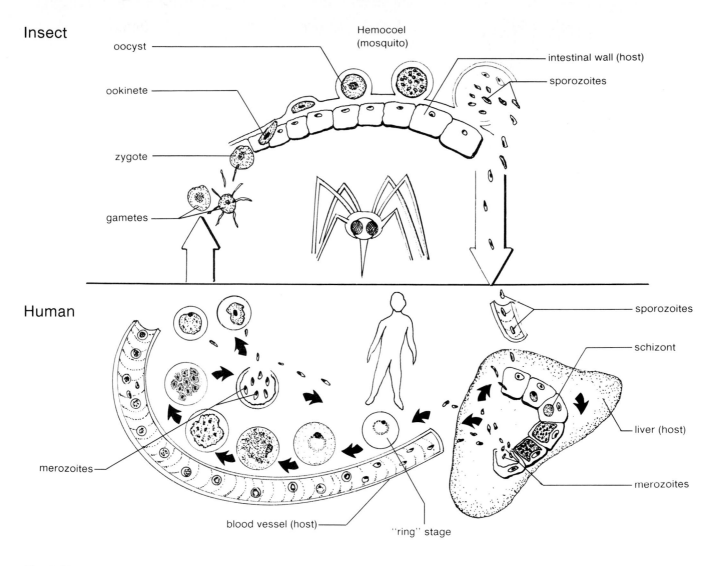

Insect

oocyst

ookinete

zygote

gametes

Hemocoel (mosquito)

intestinal wall (host)

sporozoites

Human

sporozoites

schizont

liver (host)

merozoites

merozoites

blood vessel (host)

"ring" stage

Fig. 5.24 *Life cycle of* Plasmodium.

macrogametes. The microgametocytes develop flagella-like outgrowths, which break free, become motile, and fertilize macrogametes. The fertilized macrogametes, or zygotes, then invade the gut wall of the mosquito to start the cycle.

■ Study the microscope slides with blood smears prepared with blood from humans infected with *Plasmodium vivax* or a similar species. Identify as many stages in the life cycle of *Plasmodium* as you can from your own slides and the demonstration slides. Observe the changes in morphology of the parasite during its development in the human erythrocytes. With the aid of figure 5.24, try to relate the portion of the *Plasmodium* life cycle in human erythrocytes to the other parts of the life cycle completed in the mosquito and in the exoerythrocytic stages in the human. Why do you think *Plasmodium* requires two hosts to complete its life cycle? What special adaptations for life as a parasite can you identify in *Plasmodium*? List several of these adaptations for parasitism in table 5.1.

uninfected red blood cell

ring stage

Fig. 5.25 Plasmodium. *Ring stage inside human erythrocyte. (Courtesy of Carolina Biological Supply Company, Burlington, N.C.)*

— uninfected blood cell

— macrogametocyte

Fig. 5.26 Plasmodium. *Developing macrogametocyte in human erythrocyte. (Courtesy of Carolina Biological Supply Company, Burlington, N.C.)*

Table 5.1 Adaptations for Parasitism Exhibited by *Plasmodium*

Demonstrations

1. Microscope slides with selected stages in the life cycle of *Plasmodium*.
2. Chart illustrating the life cycle of *Plasmodium*.
3. Microscope slide of *Eimeria*.
4. Chart illustrating the life cycle of *Eimeria*.

Key Terms

Cilia cylindrical cytoplasmic extensions from the surface of certain protozoa and of some metazoan cells. Serve in locomotion and feeding of ciliated protozoa. Basically similar to flagella, but shorter and more numerous. All have the universal internal 9 + 2 pattern of microtubules. Singular: cilium.

Coccidiosis disease of birds, rabbits, and other animals caused by members of the protozoon genus *Eimeria*.

Conjugation a specialized type of mating, nuclear exchange, and nuclear reorganization characteristic of ciliates. A form of sexual reproduction.

Contractile vacuole an organelle found in many freshwater protozoons that serves in osmoregulation (water balance).

Cytostome the "cell mouth" found in many protozoa, including ciliates, some flagellates, and some apicomplexa. In ciliates, the cytostome is often surrounded by specialized ciliary feeding organelles.

Flagella cylindrical cytoplasmic extensions from the surface of certain protozoa and some metazoan cells. Function in locomotion and feeding of mastigophorans. Similar to cilia but longer and usually fewer per cell. Singular: flagellum.

Gonidia specialized reproductive cells in *Volvox*. Singular: gonidium.

Macronucleus the large metabolic nucleus typical of ciliates. Often has a characteristic shape. Divides amitotically by pinching in two. Contains many duplicated sets of genes (polyploid).

Malaria disease of humans and other animals. Caused by members of the protozoon genus *Plasmodium*, which invade the blood and other tissues of the hosts.

Micronucleus the small reproductive nucleus in ciliates. Some ciliates have more than one micronucleus. Usually divides by ordinary mitosis.

Nuclear dimorphism having two distinct types of nuclei in the same cell. Characteristic of the ciliates (e.g., with macronucleus and micronucleus).

Pseudopodia the protoplasmic extensions, "false feet," of the Sarcodina used for locomotion and in feeding. Various types of sarcodines have pseudopodia specialized for certain purposes or modes of life. Singular: pseudopodium.

Stigma a light-sensitive spot found in certain flagellated protozoa, such as *Euglena* and *Volvox*.

6

Porifera

Objectives

After completing the laboratory work in this chapter, you should be able to perform the following tasks:

1. Briefly characterize the Phylum Porifera.
2. Describe the basic organization of the sponge body.
3. Explain the skeletal elements of sponges and their composition.
4. List and distinguish the three classes of sponges.
5. Differentiate between ascon, sycon, and leucon body types.
6. Explain the importance of choanocytes and describe where they are found in asconoid, syconoid, and leuconoid sponges.
7. Describe the pattern of water flow in a syconoid sponge and explain its importance.
8. Explain the role of gemmules in sponges.
9. Describe an experiment that demonstrates the ability of certain sponges to reorganize from dissociated sponge cells. Explain the significance of this experiment.

Introduction

The Phylum Porifera includes the sponges, a group of peculiar sedentary animals so different from other types of animals that they were long thought to be plants. Sponges are among the most primitive multicellular animals. They have a simple type of body organization, with a porous body permeated by a system of water canals through which water is pumped by the action of special flagellated cells (choanocytes) (figure 6.1). The body of a sponge consists of a network of epithelial cells, canals, and chambers embedded in a loose connective tissue. The only distinct tissues are the epithelial (lining) layers made up of pinacocytes and choanocytes.

Between the inner and outer layers lies the mesohyl, a loose, gelatinous matrix with several types of amoeboid cells, fibrils, and skeletal elements (spicules and/or spongin fibers). The mesohyl (formerly called mesenchyme) somewhat resembles a very loose connective tissue.

An inner layer of flagellated choanocytes (collar cells) generates water currents through the internal canal systems. These water currents are essential in the life of the sponge, for they carry food particles and oxygenated water into the sponge and waste products as well as gametes and/or larvae out of the sponge.

Sponges are sedentary filter feeders, able to capture tiny food particles measuring from about 0.1 to 50 μm from the seawater. The capture of food, which consists chiefly of fine, suspended organic particles and tiny planktonic organisms, is accomplished mainly by the choanocytes and internal amoebocytes. All digestion is intracellular (within individual cells).

Sponges lack organized multicellular organs, including nerves and muscles. In one sense, the water canal system of sponges is like an externalized circulatory system, but all the life processes of sponges take place in individual cells or groups of cells. Because of their lack of differentiated organs and because of their various other morphological and developmental peculiarities, sponges are generally believed to represent an early offshoot from the main line of animal evolution and not to be closely related to any more advanced animal types.

Of the more than 5,000 described species of sponges, most are marine, except for approximately 150 species found in freshwater streams, ponds, and lakes. Formerly, certain types of marine sponges were of considerable economic importance, and sponge fishing industries flourished in several areas, such as the Gulf of Mexico, the Caribbean Sea, and the Mediterranean Sea, where the warm, shallow waters and rocky bottoms were favorable for the growth of bath sponges (figure 6.1). Overfishing

Fig. 6.1 *Bath sponge, Class Demospongiae. One-half natural size. (Courtesy of Carolina Biological Supply Company, Burlington, N.C.)*

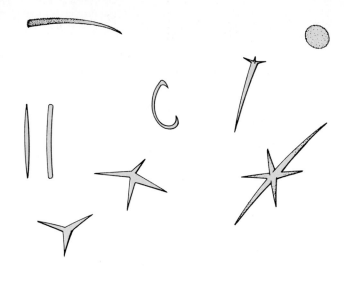

Fig. 6.2 *Examples of spicule shapes.*

and sponge diseases have taken their toll, however, and increasing competition from synthetic sponges has further diminished the commercial importance of natural sponges.

Skeleton

The skeleton of sponges is relatively complex in comparison to the general organization of the body of sponges. The skeleton is internal and consists of individual mineralized elements, called **spicules,** made up of calcium carbonate, silicon salts, and/or a network of organic fibers composed of fibrillar collagen and/or spongin fibers.

Spicules occur in a variety of forms (figures 6.2 and 6.3) and are important in the classification of sponges. Spongin is a tough, fibrous protein chemically similar to collagen, but unique to the sponges.

Classification

The division of the phylum into classes is based largely on the nature of the skeleton. Three classes of living sponges are currently recognized, the **Demospongiae,** the **Calcarea,** and the **Hexactinellida.** A new group of deep water

sponges called the coraline sponges was discovered in recent years off the coast of Jamaica. These sponges have an unusual, coral-like skeleton with a calcareous matrix with siliceous spicules and spongin fibers, and are considered by some scientists to represent a fourth class, the Sclerospongiae. Other sponge biologists, however, now include the sclerosponges in the Class Demospongiae.

Class Demospongiae

Horny Sponges. Members of this class possess a skeleton made up of a network of spongin fibers (a structural protein secreted by certain sponge cells), siliceous spicules, both, or neither. Most members of this class are marine, but two families are found in freshwater streams, ponds, and lakes. All commercial sponges belong to this class. Examples: *Spongia, Haliciona, Microciona,* and the deep sea sclerosponges (all marine); *Spongilla* (freshwater).

Class Calcarea (Calcispongiae)

Calcareous Sponges. Sponges with a skeleton consisting of many small spicules made of calcium carbonate embedded in a loose, jellylike matrix. All species are marine. Examples: *Scypha, Leucosolenia.*

Class Hexactinellida (Hyalospongiae)

Glass Sponges. Sponges with a skeleton composed of siliceous spicules, usually with six rays, as the class name implies. The spicules are often fused together into a continuous network. The tissues of glass sponges are a syncytial network of fused amoeboid cells. All glass sponges

Fig. 6.3 *Spicules from a freshwater sponge. Magnification 7,000×. (Scanning electron micrograph by F. W. Harrison.)*

are marine, and most species are found in deep areas of the world oceans. Little is known about their biology. Examples: *Euplectella* (Venus's flower basket) and *Hyalonema*.

Body Organization

Morphologically, the bodies of sponges exhibit three distinct types based on the organization of their internal canal systems. These three morphological types are designated as the **ascon, sycon,** and **leucon** types (figures 6.4 a, b, and c). It is important to recognize that these three body forms

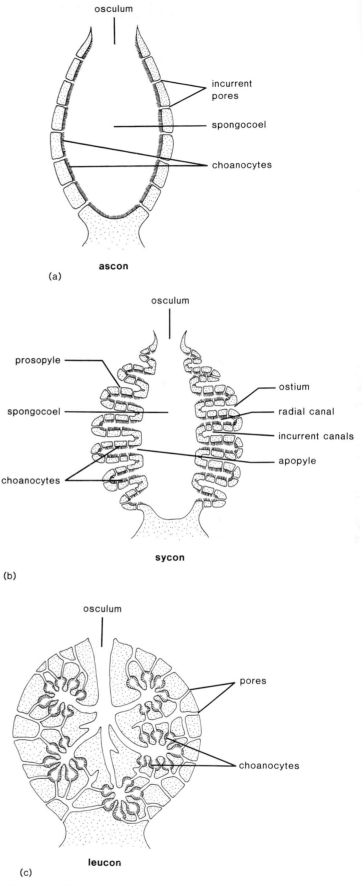

Fig. 6.4 *(a) Ascon body type, (b) sycon body type, and (c) leucon body type.*

4. The terminal opening, or **osculum,** at the upper end of each sponge. Water passes out of the sponge through this opening.

5. The **spongocoel,** a large central cavity within the sponge. This cavity is lined by the specialized, flagellated collar cells (**choanocytes**) which create water currents within the sponge.

6. Water enters the sponge through many tiny pores that penetrate the body wall.

7. Numerous triradiate (3-rayed) **spicules** may be seen embedded in the wall.

▪ In figure 6.5, make a two-inch drawing of *Leucosolenia.*

Sycon-type Sponge

Scypha (figures 6.6 and 6.7), formerly also called *Sycon* and *Grantia,* is a small, slender sponge of the sycon type.

▪ Obtain a preserved specimen, and note the size and shape of the body. Examine specimens cut in half longitudinally and observe the spongocoel, the radial canals leading from the spongocoel into the body wall, and the short collar region which leads to the osculum surrounded by a funnel of giant spicules. On the surface, observe the numerous cortical spicules that extend outward through the epidermal layer which covers the external surface of the sponge.

▪ Study a stained cross section of *Scypha* and identify the **spongocoel,** the **radial canals,** and the **incurrent canals,** which open to the exterior of the sponge through small incurrent pores.

The openings between the incurrent and excurrent canals are the **prosopyles;** the excurrent canals empty into the spongocoel through the **apopyles.** Much of the tissue within the sponge body consists of a loosely packed mesenchyme. Details of the cross section of *Scypha* are illustrated in figure 6.7.

The outer surface of the sponge is covered by a layer of thin, flat cells, called **pinacocytes.** A similar layer of pinacocytes lines the spongocoel. The **choanocytes** (figure 6.8) are found lining the radial canals, which empty into the spongocoel through the apopyle (figure 6.9). These choanocytes are small and difficult to identify in most microscopic preparations. You should be able to find some

Fig. 6.5 *Student drawing of* Leucosolenia.

represent morphological types and are not directly related to the three classes of sponges. In fact, only a few species of sponges exhibit the ascon and sycon body types. The majority of sponges are of the leucon type.

Ascon-type Sponge

Leucosolenia is a small colonial sponge of the ascon type. Examine a preserved specimen or a microscopic whole mount with your stereoscopic microscope and observe the following features.

1. A system of horizontal tubes that bear numerous upright branches.

2. The upright branches that represent individual sponges of the colony.

3. Buds formed on the sides of the individual sponges.

Fig. 6.6 Scypha, *typical cluster of sponges. (Courtesy of Carolina Biological Supply Company, Burlington, N.C.)*

large undifferentiated amoeboid cells within the body wall, which represent the **mesohyl.** Eggs and developing embryos are also found within the body wall. There are no differentiated sex organs.

Leucon-type Sponge

Leuconoid sponges are structurally the most complex and also the most common body type among the living sponges. All freshwater sponges and most marine sponges are leuconoid.

■ Study a portion of a preserved specimen of a bath sponge and note its rubbery texture and the complex system of branching canals.

Attached to the radial canals are numerous small, spherical, flagellated chambers. Collar cells are found lining only these tiny flagellated chambers in leuconoid sponges. Study also a dried specimen of a bath sponge and note how its texture differs from that of the preserved specimen. Only the network of **spongin fibers** remains in the dried sponge. Observe the microscopic demonstration slide of spongin fibers.

1. Eggs and developing embryos of *Scypha* (microscope slide).
2. Assorted spicules (microscope slide).
3. Spongin fibers (microscope slide).
4. Glass sponge and examples of other types of sponges.
5. Skeletons of glass sponges.
6. Preserved and dried samples of bath sponges and other types of sponges.

Freshwater Sponges

Although most sponges are marine, a few species live in freshwater streams, ponds, and lakes. Freshwater sponges are much less prominent than their ubiquitous marine relatives. The freshwater sponges grow as tufts or small irregular masses encrusting sticks, stones, or submerged plants. Most species are yellow or brown in color, but a few species are green because of symbiotic algae that live within the sponge.

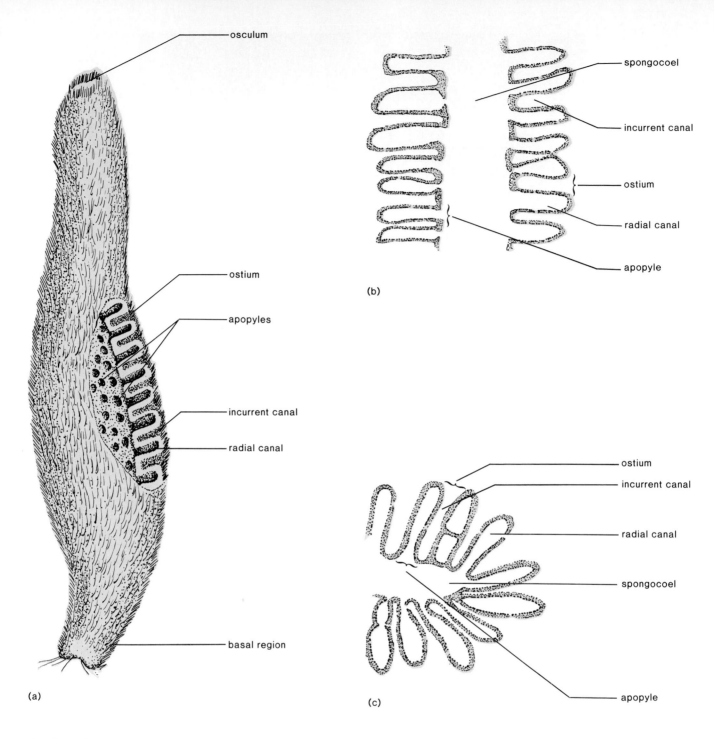

Fig. 6.7 Scypha: *(a) external view with portion of body wall removed to show interior detail; (b) longitudinal section; (c) cross section.*

Chapter 6

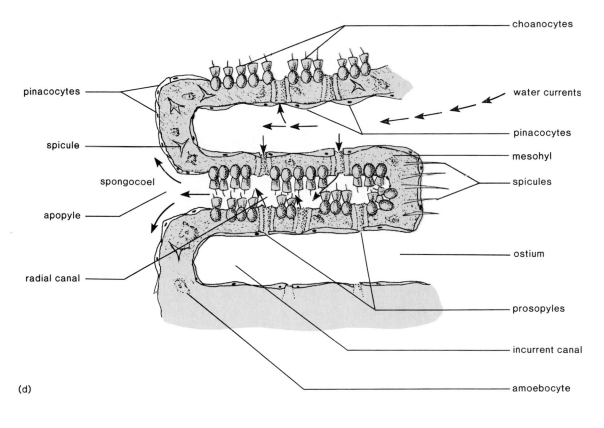

pinacocytes

spicule

spongocoel

apopyle

radial canal

choanocytes

water currents

pinacocytes

mesohyl

spicules

ostium

prosopyles

incurrent canal

amoebocyte

(d)

Fig. 6.7 *Continued. (d) detail of radial and incurrent canals, and pattern of water currents.*

collar of choanocyte

flagellum

body of choanocyte

water canal

Fig. 6.8 *Cross section through choanocytes of* Ephydatia, *a freshwater sponge. Magnification 21,000×. (Scanning electron micrograph of a freeze fracture section by Louis de Vos.)*

All freshwater sponges (and a few marine sponges) form internal asexual buds called **gemmules** (figure 6.10). Gemmules are resistant stages which serve to carry the sponge through the winter or aid in survival during a drought. Observe the microscope slide of gemmules on demonstration. What important features of the gemmule aid in the survival of the species?

Regeneration and Reconstitution (Optional Exercise)

Sponges are especially noted for their powers of regeneration. A small part of a sponge can regenerate a complete sponge. Many years ago, H. V. Wilson showed that pieces of a living sponge pressed through a fine cloth mesh to separate the sponge into individual cells and clusters of cells could reassemble and develop into a complete new sponge. More recent experiments with the separation and reassociation of sponge cells have provided important information about the nature and mechanism of cell recognition processes and the organization of tissues during development.

If living sponges are available in your laboratory, you may be able to repeat this famous experiment. *Microciona,* a common red sponge on the Atlantic coast of the United States, was used in the original experiments. You will need a small finger bowl, a watch glass, a clean microscope slide, a pipette, some silk mesh cloth, and some seawater for this experiment.

Procedure
1. Fill the finger bowl about two-thirds full of seawater of the proper ionic strength (depending on the source of the sponges) and place the watch glass on the bottom of the dish. Place the microscope slide on top of the watch glass.

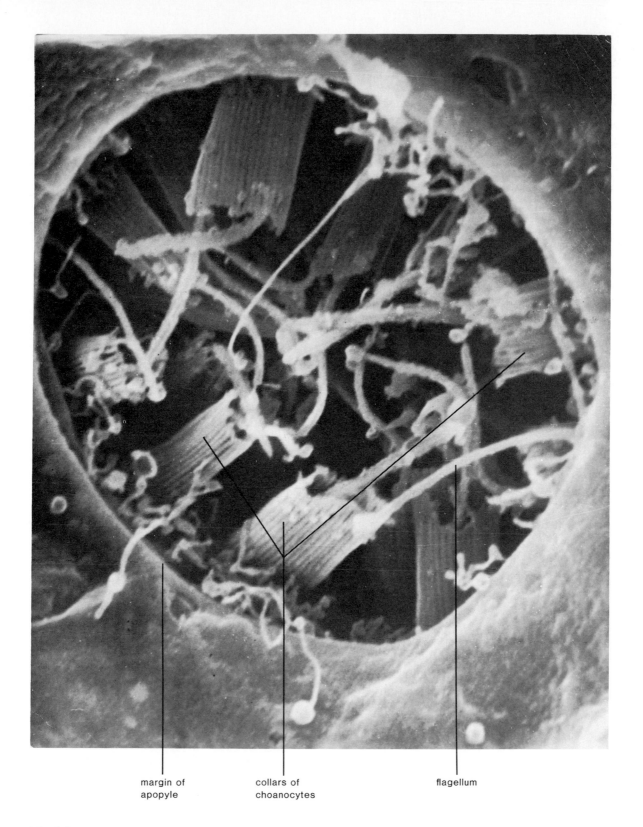

margin of
apopyle

collars of
choanocytes

flagellum

Fig. 6.9 *Opening of apopyle showing arrangement of choanocytes within. Magnification 21,000×. (Scanning electron micrograph by Louis de Vos.)*

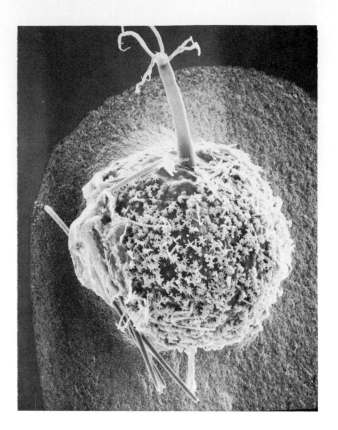

Fig. 6.10 *Gemmule of freshwater sponge. Many specialized spicules are embedded in the resistant covering of the spicule. Magnification 16,000×. (Scanning electron micrograph by F. W. Harrison.)*

2. In a separate bowl of seawater, prepare a suspension of cells and fragments of *Microciona* by pressing small pieces of sponge through a fine silk bolting cloth or nylon mesh of about 150–200 micrometers in diameter.

3. Pipette a small amount of this cell suspension onto the slide and allow the cells to settle.

4. Carefully lift out the slide and observe the cells. Make similar observations at intervals during the next 24–48 hours. Watch for the initiation of cellular aggregation and the thin protoplasmic extensions (filopodia) put out by the small clumps of aggregating cells. Sketch the cells and clumps of cells as a record of your observations.

Demonstrations

1. Gemmules (microscope slide).
2. Living or preserved freshwater sponges.

Key Terms

Ascon the simplest type of sponge body form with a central spongocoel lined with choanocytes and with incurrent pores opening directly into the spongocoel.

Choanocyte a special type of flagellated collar cell characteristic of sponges that lines the flagellated water chambers and canals.

Gemmule a dormant stage in the life cycle of freshwater sponges and a few species of marine sponges. Formed as internal asexual buds, they aid the sponges in surviving adverse environmental conditions.

Leucon a complex type of sponge body form with an intricate system of internal water canals. Choanocytes line certain small chambers within the system.

Mesohyl the loose, gelatinous matrix in a sponge body containing several types of amoeboid cells, fibrils, and skeletal elements (spicules and/or spongin fibers). Located between the outer pinacoderm and the inner choanocyte layer. Formerly called mesenchyme.

Pinacocyte a thin, flattened type of cell that lines the outer surface of sponges and most inner surfaces not lined by choanocytes.

Pinacoderm outer lining of sponge body made up of pinacocytes.

Spicules individual skeletal elements that make up the skeleton of most sponges. Consist mainly of calcium carbonate or silicon salts and exhibit many different shapes.

Spongin fibers a loose network of protein fibers that form part or all of the skeleton in most horny sponges (Class Demospongia).

Sycon a type of sponge body that exhibits a central spongocoel into which numerous radial canals lined with choanocytes empty.

7

Cnidaria

Objectives

After completing the laboratory work in this chapter, you should be able to perform the following tasks:

1. List the three classes of cnidarians and briefly characterize each class.
2. Describe the general morphology of *Hydra*.
3. Identify the main morphological features of *Hydra* on a microscope slide or illustration.
4. Discuss the structure and function of nematocysts and explain their importance in *Hydra* and other cnidarians.
5. Describe the feeding reaction of *Hydra* and its control.
6. Describe the reproductive processes of *Hydra* and be able to identify mature male and female individuals.
7. Identify the chief morphological features of the medusa of *Gonionemus* and explain the function of each part.
8. Describe the life cycle of *Gonionemus*.
9. Identify the main morphological features of a mature *Obelia* colony and explain the function of each.
10. Describe the life cycle of *Obelia* and explain how it illustrates the alternation of generations commonly found in cnidarians.
11. Describe the main morphological features of the scyphozoan medusa *Aurelia* and explain the function of each.
12. Describe the life cycle of *Aurelia*.
13. Describe the structure of the sea anemone *Metridium* and explain the function of its principal parts.

Introduction

Cnidarians (formerly called coelenterates) are the simplest animals with definite tissues; the cnidarian body consists of two well-defined tissue layers and a third intervening layer of gelatinous material, the **mesoglea,** that varies in structure among the three classes of cnidarians. The outer **epidermis** layer covers the external surface of the body, and the inner **gastrodermis** layer lines a single internal body cavity. In the simplest cnidarians, Class Hydrozoa, the mesoglea is thin and gelatinous; in the Class Scyphozoa, the mesoglea is fibrous and contains amoeboid cells; and in the Class Anthozoa, the mesoglea has many amoeboid cells and thus resembles a mesenchyme layer.

The cnidarians and the related Phylum Ctenophora are especially noted for their prominent **radial symmetry.** For this reason, these two groups are commonly referred to as the radiate phyla.

The name Cnidaria is derived from the **cnidocytes,** special cells which produce the nematocysts or "stinging capsules" characteristic of this phylum. The alternate phylum name, Coelenterata, refers to the large **coelenteron,** or gastrovascular cavity, also characteristic of the group. The related Phylum Ctenophora derives its name, meaning "comb-bearer," from its characteristic eight rows of **comb plates,** or ctenes, which serve for locomotion.

Two basic body forms are exhibited by the Cnidaria: an attached **polyp** stage and a free-swimming **medusa** stage. Many species exhibit both a polyp stage and a medusa stage, and their life cycles involve an **alternation** of these two body forms or "generations." Two other important distinguishing characteristics of the phylum include **tentacles** around the mouth and a diffuse **nerve net,** which provides a modest degree of nervous coordination.

Classification

The phylum is divided into three classes:

Class Hydrozoa (Hydroids and Siphonophores)

Animals usually have both polyp and medusa in the life cycle, medusae with a velum, and gonads on radial canals of medusae. Mesoglea is thin and gelatinous; well-defined epidermis and gastrodermis. Freshwater and marine species. Examples: *Obelia*, *Hydra*, *Gonionemus*, and *Physalia* (Portuguese man-of-war).

Class Scyphozoa (True Jellyfish)

Mainly large marine jellyfish with abundant mesoglea, polyp stage reduced or absent, no velum in medusae. Mesoglea containing fibers and amoeboid cells. All marine. Examples: *Aurelia*, *Chrysaora* (sea nettle), and *Cyanea*.

Class Anthozoa (Sea Anemones and Corals)

Solitary or colonial animals with polyp stage only, medusa absent, pharynx or gullet present, gastrovascular cavity partitioned by septa. Mesoglea with many amoeboid cells. All marine. Examples: *Metridium* (sea anemone), *Astrangia* (coral), *Gorgonia* (sea fan), and *Renilla* (sea pansy).

Materials List

Living specimens
  Hydra
  Daphnia or *Artemia* larvae (food for *Hydra*)
  Sea anemones in aquarium (Demonstration)

Preserved specimens
  Gonionemus, medusa
  Metridium
  Physalia
  Aurelia, medusae
  Ctenophores (Demonstration)
  Representative Hydrozoa, Anthozoa, and Scyphozoa (Demonstrations)
  Representative anemones and corals (Demonstration)

Prepared microscope slides
  Hydra, cross section, longitudinal section
  Obelia, whole mount of polyp, whole mount of medusa
  Nematocysts, discharged (Demonstration)
  Gonionemus, statocysts (Demonstration)
  Hydra, male with testes, whole mount (Demonstration)
  Hydra, female with ovary, whole mount (Demonstration)
  Green *Hydra* with symbiotic algae, whole mount (Demonstration)
  Planula larva, whole mount (Demonstration)
  Aurelia marginal sense organs, scyphistoma, ephyra larva, strobila (Demonstrations)

Chemicals
  1% Acetic acid
  0.01% Methylene blue solution
  Pond water

A Polyp: *Hydra*

Class Hydrozoa

Hydra (figure 7.1) typifies the polyp form of a cnidarian. *Hydra* lives in freshwater streams, lakes, and ponds, where it is usually attached to submerged sticks, stones, or vegetation and feeds on various small aquatic animals. About fourteen species of *Hydra* are known to occur in the United States.

Although *Hydra* serves as a good example of the polyp form of a cnidarian, it differs in several important ways from most members of the Class Hydrozoa: (1) *Hydra* is solitary, while most hydrozoan polyps are colonial; (2) *Hydra* has no separate medusa stage, which most hydrozoan species have; (3) the polyp of *Hydra* bears gonads, unlike most hydrozoan polyps; (4) *Hydra* is

mobile, unlike most hydrozoan polyps, which are sessile; and (5) *Hydra* lives in fresh water, while most hydrozoans are marine.

General Appearance and Morphology

■ Examine a living specimen of *Hydra* in a dish of pond water. Caution: Be sure to use pond water and not tap water, since most tap water contains trace amounts of copper and other substances toxic to *Hydra*.

Locate the following structures on your specimen: (1) the **basal disc** at the lower end which serves for attachment; (2) the cylindrical **body;** (3) a circle of **tentacles** at the free end (how many?); (4) the **hypostome,** an elevation between the bases of the radially arranged tentacles; (5) the **mouth** in the center of the hypostome; (6) **buds,** the products of asexual reproduction, may also be present; and (7) **ovaries** or **testes** are also present on the lower portion of the body in mature specimens (figure 7.2).

Behavior

■ Does your specimen change in shape? Touch one of the tentacles with the tip of your dissecting needle. What is the reaction? What methods of locomotion are used by *Hydra?* Observe *Hydra* feeding in an aquarium, or add a few *Daphnia* or washed brine shrimp (*Artemia*) larvae to your dish near the specimen to observe feeding.

Cnidocytes and Nematocysts

■ After you have studied the basic form and behavior of your specimen, place it on a clean microscope slide in a drop of water and carefully add a cover glass. Observe the numerous **cnidocytes,** which appear as swellings on the tentacles. Each cnidocyte is a cell containing a **nematocyst,** or stinging capsule.

The cnidocytes of *Hydra* can be stained to aid in your observations by adding a drop of 0.01% methylene blue solution to the edge of the coverslip.
  Hydra has four different kinds of **nematocysts,** each with a distinctive structure and function (figure 7.3). Nematocysts are important in the capture of food, in locomotion, and in attachment. Nematocysts are complex **secretory products** formed within developing cnidocytes. Each nematocyst consists of an outer protein **capsule,** and a long, coiled **tube** often armed with **spines** or **barbs.** Toxins, enzymes, and other chemicals are also contained within the capsule of certain kinds of nematocysts.
  Tap on the coverslip of your wet mount of *Hydra* tentacles to induce discharge of the nematocysts. When properly stimulated, nematocysts empty their contents with a rapid discharge as the coiled tube is shot out. In addition to tapping the coverslip to induce nematocyst

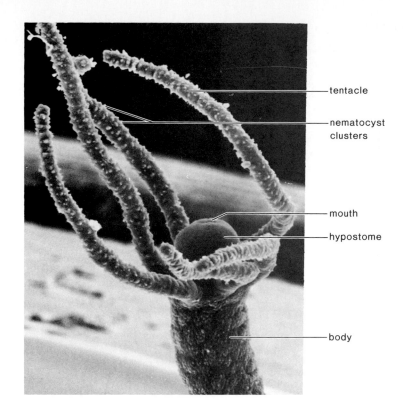

Fig. 7.1 Hydra. *(Scanning electron micrograph by Louis de Vos.)*

discharge, you may also wish to try adding a drop of dilute acetic acid (1% solution) to stimulate nematocyst discharge.

Study the discharged nematocysts under high power or with an oil immersion lens, if available. Observe the outer capsule, the long thread or tube, and the large spines or barbs at the base of the tube. How many kinds of nematocysts can you find on your slide?

Observe also the microscopic demonstration of discharged nematocysts.

Histological Structure

Hydra and other members of the Class Hydrozoa exhibit the simplest histological structure among the Cnidaria (see figure 7.2).

■ Examine stained slides of cross and longitudinal sections of *Hydra,* and note the following: (1) **epidermis,** the outer, thinner epithelial layer of cells; (2) **gastrodermis,** the inner layer of cells; (3) **mesoglea,** a very thin, noncellular layer between the epidermis and gastrodermis; and (4) **gastrovascular cavity,** or enteron, the internal cavity lined by the gastrodermis.

Cellular Structure

Hydra shows distinct advances compared with a sponge, not only in its general structure but also in the degree of differentiation of cellular structure and functions. Several different types of cells may be distinguished in the stained preparations. In the epidermis, under high power, try to distinguish the following cell types: (1) The large and abundant **epitheliomuscular cells,** which possess contractile processes or fibers at their base, all running lengthwise. What specific functions do these fibers perform? (2) The small **interstitial cells** at the bases of the epitheliomuscular cells. (3) The **cnidocytes.** (4) The mucus-secreting **gland cells** abundant on the pedal disc. Also among the epidermal cells are many small nerve cells as described in the next section. They can be seen only in specially stained slides.

In the inner gastrodermal layer, note the following cell types: (1) The abundant, large, vacuolated **digestive cells.** These cells bear one or two flagella, and ingest food particles for intracellular digestion. They are also epitheliomuscular in character and possess contractile fibers that run transversely along their bases, thus providing a circular musculature. What, therefore, is the function of the contractile fibers in the gastrodermal cells? Compare this with the function of the fibers in the epidermis. (2) The **gland cells,** which secrete either mucus or digestive enzymes. Mucus-secreting cells are abundant in the hypostome region. (3) The **interstitial cells** at the bases of the gastrodermal cells. Several other types of cells are also present but are difficult to observe except in specially stained preparations.

Nervous System

Specialized nerve cells, or **neurons,** are located among the cells of both the epidermal and gastrodermal layers of *Hydra.* These neurons may have two or more processes that connect via **synaptic junctions** with other neurons or with various types of receptor or effector cells.

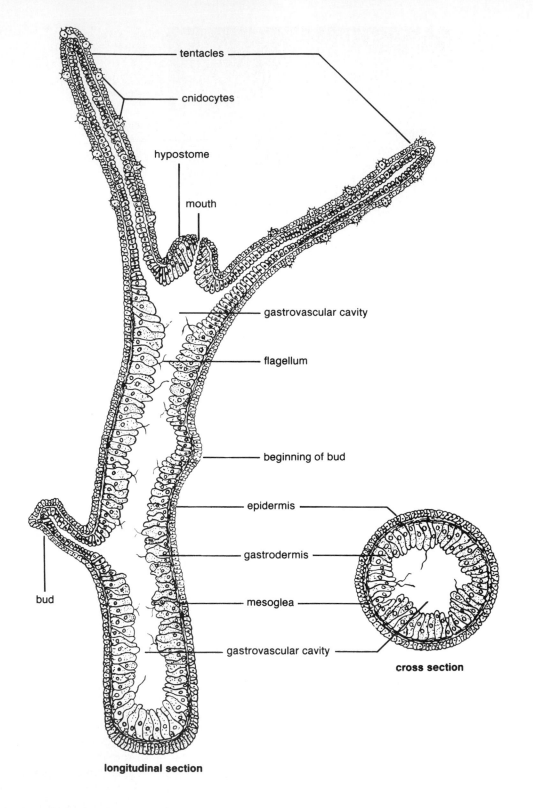

tentacles

cnidocytes

hypostome

mouth

gastrovascular cavity

flagellum

beginning of bud

epidermis

gastrodermis

mesoglea

gastrovascular cavity

bud

cross section

longitudinal section

Fig. 7.2   Hydra. *Longitudinal and cross sections.*

Fig. 7.3 Hydra. *Nematocyst types. (Courtesy of Carolina Biological Supply Company, Burlington, N.C.)*

Figure 7.4 shows an insolated multipolar neuron from the body wall of *Hydra*. Special techniques are required to see cnidarian neurons with either a light or an electron microscope.

The interconnecting network of neurons in *Hydra* form a **nerve net** that lacks concentrations of neurons that form ganglia or a brain. This diffuse nerve net is characteristic of *Hydra* and other cnidarians. Nervous impulses tend to spread in a radiating pattern from the point of origin or stimulation because of the structure of synapses between adjacent neurons. Many of these synapses are symmetrical and are nonpolarized, thus allowing impulses to flow in both directions. Some polarized synapses also occur in cnidarians. Recent studies have suggested that remnants of the diffuse nerve net of cnidarians may be represented in the nervous system of certain higher animals, including the digestive systems of annelid worms and of humans. The rhythmic peristaltic movements of your stomach and intestine after you eat are coordinated by a similar nerve net.

Neurons also have processes connecting via synapses with epithelial muscular cells, gland cells, and nematocysts. Substantial evidence now suggests that nematocyst discharge is at least partly under nervous control.

Feeding Behavior

Hydra is a carnivore and feeds on living crustaceans, rotifers, insect larvae, and other small animals (figure 7.5). When properly stimulated, *Hydra* exhibits a characteristic feeding response. You can observe the feeding behavior by placing a healthy, unfed *Hydra* in a clean watch glass containing about 10 ml of pond water (or *Hydra* culture solution).

Fig. 7.4 *Isolated multipolar neuron from body column of* Hydra. *(Scanning electron micrograph by J. A. Westfall and L. G. Epp from* Tissue and Cell *17 no. 2:165.)*

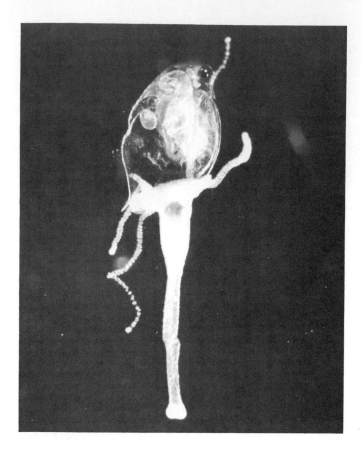

Fig. 7.5 Hydra *ingesting a* Daphnia. *(Courtesy of Carolina Biological Supply Company, Burlington, N.C.)*

■ Add a few (6–12) washed brine shrimp (*Artemia*) larvae near the *Hydra* and record your observations. Note how the food organisms are captured, what happens to them after their capture, and the movements of various parts of the *Hydra*. Use the second hand on your watch (or your neighbor's watch!) to time various parts of the complex behavioral reaction of the *Hydra*. Make a record of your observations on the pages provided for *Notes and Sketches* at the end of this chapter.

Scientists have shown that the feeding reaction of *Hydra* is normally caused by body fluids oozing from the body of its prey, which has been pierced by nematocysts. Further experiments have shown that a feeding reaction can also be elicited by a solution of a tripeptide, reduced glutathione.

Reproduction

Hydra reproduces asexually by budding and sexually by the production of eggs and sperm (figure 7.6). The hydra in most laboratory samples are asexual, and you can often find budding specimens among them. In nature, hydra reproduce most of the year by budding and at certain times form gonads, testes, or ovaries, which appear as thickenings of the body wall (figure 7.7). Most species of *Hydra* are **dioecious** (sexes separate) although a few species are **monoecious** (both sexes in one individual). Observe the demonstration materials of male and female hydra with gonads.

Regeneration (Optional Exercise)

Hydra and other cnidarians have great powers of regeneration. Place a *Hydra* in a watch glass and cut across the middle of the body with a sharp scalpel or razor blade. Separate the two pieces into different watch glasses and observe their development during the next several days. Observe the formation of a new hypostome, mouth, and tentacles by the basal half of your *Hydra*. What happens to the other half? Keep a record of your observations in the pages provided for *Notes and Sketches* at the end of this chapter.

Demonstrations

1. *Hydra* with testes (microscope slide).
2. *Hydra* with ovary (microscope slide).
3. Symbiotic algae in green hydra, *Chlorohydra* (living or microscope slide).

A Medusa: *Gonionemus*

Class Hydrozoa

Gonionemus is a small marine medusa, or jellyfish, common in many parts of the world. The medusa of *Gonionemus* (figure 7.8) is the adult or sexually mature stage and serves to illustrate the typical structure of a cnidarian medusa (figure 7.9).

■ Examine a preserved specimen in a watch glass partly filled with water and note its umbrellalike form. The specimens are delicate and must be handled with care. Note the jellylike consistency of the medusa, due to the thick layer of **mesoglea** within the body. Most of the bulk of the medusoid body consists of mesoglea.

The medusa usually swims with its convex **exumbrellar surface** upward and the concave **subumbrellar surface** downward. Observe the numerous **tentacles** around the margin of the bell. Extending downward from the center of the subumbrellar surface is the **manubrium,** with the **mouth** at its tip surrounded by four **oral lobes.** Around the inner margin of the bell, extending inward, is a thin circular flap of tissue, the **velum,** which is believed to aid

Fig. 7.6 Hydra *with bud and fertilized egg attached. (Courtesy of Carolina Biological Supply Company, Burlington, N.C.)*

in swimming. At the base of the manubrium is an expanded portion of the manubrium, the "stomach." Find the four **radial canals** extending from the stomach to the **circular canal** at the margin of the bell. The hollow tentacles connect with the circular canal, and their cavities thus represent a continuation of the gastrovascular cavity. Observe the numerous **nematocyst batteries** and the **adhesive pads** on the tentacles.

At the bases of the tentacles are round, pigmented structures believed to be light sensitive photoreceptors. Between the tentacle bases are the **statocysts,** which serve as balancing organs.

Observe the folded gonads attached to the subumbrellar surface of the radial canals. The gonads on the specimen you are studying are either ovaries or testes, since *Gonionemus,* like most cnidarians, is dioecious. Gametes produced by the gonads are released into the sea, and the fertilized eggs develop into a ciliated **planula larva.** Later, the planula larva settles and attaches to some submerged object in the sea and transforms into a microscopic **polyp stage.** The polyp stage may reproduce asexually by budding and, under certain conditions, may form tiny medusa buds that detach and grow to develop into mature medusae, thus completing the life cycle.

Demonstrations
1. Statocysts at margin of *Gonionemus* (wet mount or microscope slide).
2. Planula larva (microscope slide).

A Colonial Polyp: *Obelia*

Class Hydrozoa

Obelia is a colonial marine cnidarian which illustrates the complex life cycle with alternating polyp and medusa stages found in many cnidarians (figure 7.10).

■ Examine a stained whole mount of the polyp or asexual stage of *Obelia* and study its organization. Like many colonial animals, *Obelia* exhibits **polymorphism** or morphological specialization of its members.

Therefore you can distinguish two different kinds of individuals in an *Obelia* colony, feeding polyps or **hydranths** and reproductive polyps or **gonangia.** The feeding

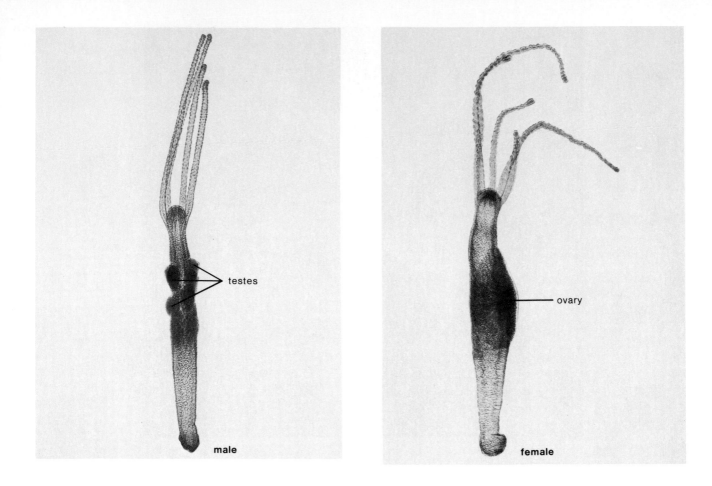

Fig. 7.7 *Male and female* Hydra *with gonads. Cross section. (Courtesy of Carolina Biological Supply Company, Burlington, N.C.)*

Fig. 7.8 Gonionemus *medusa. Magnification 3×. (Courtesy of Carolina Biological Supply Company, Burlington, N.C.)*

Fig. 7.9 Gonionemus *medusa. Vertical section*

Fig. 7.10 Obelia, *life cycle.*

polyps bear tentacles armed with nematocysts, a mouth, a hypostome, and a delicate outer covering, the **hydrotheca** (an extension of the perisarc). Reproductive polyps consist of a central **blastostyle** on which **medusa buds** develop, and a thin outer covering, the **gonotheca.** At the distal end of the gonotheca is an opening, the **gonopore,** through which the newly liberated medusae escape. Note that the gonangia have no mouth or tentacles. How do they receive their nutrition?

The hydranths and gonangia are attached to a main stem, or **hydrocaulus,** which consists of a cylindrical tube of living tissue, the **coenosarc,** and an outer secreted cov-

ering, the **perisarc.** In cross section, the coenosarc resembles a cross section of *Hydra* with an outer epidermis, an inner gastrodermis, and a thin intervening layer of mesoglea.

Alternation of Generations

The life cycle of *Obelia* is fundamentally similar to that of *Gonionemus* and illustrates the alternation of generations characteristic of the Phylum Cnidaria (figure 7.10).

The alternation of the asexual polyp generation and the sexual medusa generation (also called metagenesis) is basically different from the alternation of generations in

Fig. 7.11 *Student drawing of Obelia* medusa.

plants. Both the polyp and medusa generations of cnidarians are **diploid** (2N), while in plants the alternation is between a **haploid** (1N) gametophyte generation and a **diploid** (2N) sporophyte generation.

In the life cycle of *Obelia,* the polyp generation produces medusa buds within its gonangia. The tiny, short-lived medusae escape into the plankton and produce either eggs or sperm. Fertilized eggs develop into ciliated **planula larvae,** which swim about in the sea for a time and settle to transform into a new polyp. Buds transformed by asexual reproduction of the polyp do not detach, thus forming a colony.

The sexual stage of *Obelia* is a free-swimming medusa somewhat similar to *Gonionemus,* but much smaller in size.

■ Observe a stained whole mount of an *Obelia* medusa and observe its structure. Draw a picture of an *Obelia* medusa in figure 7.11.

Demonstration
Preserved specimens of *Tubularia, Physalia,* and/or other hydrozoans.

A Portuguese Man-of-war: *Physalia*

Class Hydrozoa

The Portuguese man-of-war (figure 7.12), *Physalia,* is a complex colonial hydrozoan (Class Hydrozoa, Order Siphonophora) exhibiting a high degree of **polymorphism.**

Fig. 7.12 Physalia, *Portuguese man-of-war. About one-half life size. (Photograph by C. F. Lytle.)*

A single colony may consist of as many as 1,000 individuals and several types of polypoid and medusoid forms. The familiar iridescent colonies of *Physalia* are commonly found along the beaches of Florida, the South Atlantic, the Gulf of Mexico, and, sometimes, as far north as Cape Cod.

The most prominent feature of *Physalia* is a gas-filled **float** above which is a sail-like **crest.** *Physalia* is transported by winds and oceanic currents, and its normal habitat is the open sea rather than the sandy beach where it is most often seen (and sometimes felt!) by bathers. The sting from the nematocysts on the tentacles of *Physalia* can be painful when touched but is rarely dangerous, except to highly sensitive individuals.

Below the float are suspended numerous **tentacles** and other structures made up of several kinds of modified polyps and medusae. Thus, *Physalia* is an unusual cnidarian since a single colony contains both polypoid and medusoid individuals of several types closely joined together, in contrast to separate polypoid and medusoid generations.

■ Observe a preserved or a plastic-embedded specimen of *Physalia* and identify the float, the crest, and the tentacles. See how many types of individuals you can identify among the structures attached to the float.

Velella and *Porpita* are two related colonial hydrozoans often seen on our Pacific coast and less often in the Gulf of Mexico and South Atlantic. Both exhibit polymorphism and habits similar to *Physalia*.

A Scyphozoan Jellyfish: *Aurelia*

Class Scyphozoa

Aurelia (figure 7.13) is a common, widely distributed marine jellyfish. Large specimens may reach 30 centimeters (12 inches) in diameter. The polyp form, called a **scyphistoma**, is small, sessile, and lives attached to rocks and other submerged objects in shallow coastal waters.

■ Study a preserved specimen of *Aurelia* to learn about the organization of a scyphozoan medusa. You will not need to dissect the specimen, since the transparent body readily shows most important features. Handle the specimen with care and return it for study by another student.

Observe the four-part **radial symmetry** and locate the four long **oral arms** arising from the corners of the square mouth. Along the arms find the many short **oral tentacles** that help to capture food (small planktonic animals), which are then moved toward the mouth along the **ciliated groove** on the oral side of each arm. After passing through the mouth, the food enters the gastrovascular cavity. Internally the gastrovascular cavity is divided into four **gastric pouches.** A ring of **gastric filaments** within each gastric pouch immobilizes or kills any food organisms still active. The gastric filaments bear many nematocysts.

Four horseshoe-shaped **gonads** surround the ring of gastric filaments within the four gastric pouches. Depressions on the subumbrellar surface of the bell beneath the gonads are called **subgenital pits.** Their function is unknown.

Observe the complex branching system of **radial canals** that distribute food materials from the gastric pockets to other parts of the bell, and an outer **circular canal** around the margin of the bell. Also, around the margin of the bell, locate the eight marginal **sense organs.** These marginal organs are sensitive organs of touch and balance.

Reproduction and Life Cycle

Mature *Aurelia* medusae release gametes from the gonads into the gastrovascular cavity. How does this differ from the hydrozoan medusae? The gametes exit from the mouth and fertilization is external.

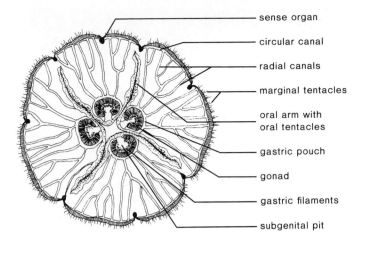

Fig. 7.13   Aurelia *medusa, oral view.*

The fertilized eggs or zygotes develop into ciliated **planula larvae,** which may be retained for a time on the oral arms of the medusa and later settle to the sea bottom. There the larvae develop into a small, trumpet-shaped polyp, called a **scyphistoma.**

Under appropriate environmental conditions, the scyphistoma transforms into a **strobila** (figure 7.14). The strobila develops and releases by transverse fission a series of saucer-shaped **ephyra larvae,** which bear marginal sense organs and other medusoid features. The ephyra larvae gradually transform into adult jellyfish to complete the life cycle.

■ Study the demonstration materials and draw a scyphistoma in figure 7.15.

Demonstrations
1. Marginal sense organ of *Aurelia* medusa (microscope slide).
2. Scyphistoma of *Aurelia* or other scyphozoan (preserved or microscope slide).
3. Planula and ephyra larvae of *Aurelia* (microscope slide).
4. Strobila stage of *Aurelia* (microscope slide).

A Sea Anemone: *Metridium*

Class Anthozoa

Sea anemones are typically sessile cnidarians that attach to rocks, shells, pilings, and other hard substrates in the sea. Some species, however, burrow in soft bottoms or are even free-swimming. All nonetheless represent the polyp form of the cnidarians; the medusa generation is totally lacking in this class. The anemones and other Anthozoa represent the highest degree of specialization of the cnidarian polyp. The basic features of the anthozoan polyp

Fig. 7.14 Aurelia *strobila. (Photograph by C. F. Lytle.)*

Fig. 7.15 *Student drawing of an* Aurelia *scyphistoma.*

are well illustrated by the common North Atlantic anemone, *Metridium* (figure 7.16). The same or a similar form also occurs in the Pacific Ocean.

▪ Select a preserved specimen of *Metridium* and identify the **mouth** in the center of the **oral disc** surrounded by many short **oral tentacles,** and the **basal disc,** which attaches to rocks or other hard substrates.

Most of the internal anatomy can be observed in specimens that have been bisected longitudinally or horizontally. It is not necessary, therefore, to dissect the specimens further. Study precut preserved specimens and locate the following structures. Find the tubular **gullet** leading internally from the mouth to the large **gastrovascular cavity.** One or more **ciliated grooves** (siphonoglyphs) should be found along the edge of the gullet. Usually the cilia within these grooves beat inward to provide a respiratory current of water into the gastrovascular cavity. Cilia on the remainder of the gullet wall beat outward to remove wastes and foreign particles from the gastrovascular cavity. During feeding, however, the beat of the cilia along the gullet wall is reversed and aids in moving food particles into the gastrovascular cavity. Observe the feeding of living anemones in an aquarium (if available) to better understand the interaction of the tentacles and ciliary currents on the oral disc and in the gullet during the feeding process.

The gastrovascular cavity is partially divided into sections by thin vertical walls of tissue called **septa.** Some septa attach to the central gullet; others (partial or incomplete septa) bear thickened **septal filaments** on their free inner margins. The septal filaments extend below the partial septa as thin, twisted, threadlike **acontia.** The acontia bear numerous nematocysts and act to subdue living prey taken into the gastrovascular cavity. Distal to the septal filaments on the partial septa, find the **gonads,** which appear as thickened ridges parallel to the septal filaments. On the septa near the oral disc, locate the **ostia,** round openings that allow circulation of fluid between adjacent sections of the gastrovascular cavity.

Reproduction and Life Cycle

Reproduction in *Metridium* and most other anemones is both asexual and sexual. Some anemones reproduce asexually by splitting longitudinally (**longitudinal fission**), but the main asexual means of reproduction in *Metridium* is **pedal laceration.** Bits of tissue from the pedal disc are split from the anemone as the animal moves along the substrate. These tissue pieces later regenerate an entire small anemone, literally in the footsteps of its parent.

Sexual reproduction occurs seasonally when gametes are released from the gonads on the partial septa into the gastrovascular cavity. The gametes are released and are fertilized in the sea. The fertilized eggs develop into free-swimming **planula larvae.** After a period as planktonic larvae, the planulae settle on some hard substrate and metamorphose into an anemone.

Corals

Class Anthozoa

Most anthozoans are corals, the largest and best known of which are the **stony** or **scleractinian corals** (Order Zoantharia) and the **octocorals** (Subclass Alcyonaria),

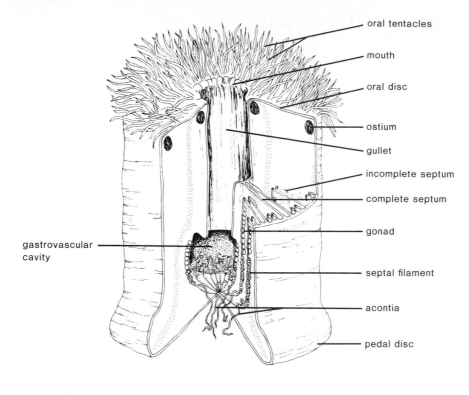

Labels (clockwise from top right): oral tentacles, mouth, oral disc, ostium, gullet, incomplete septum, complete septum, gonad, septal filament, acontia, pedal disc, gastrovascular cavity

Fig. 7.16 Metridium, *partly dissected.*

which includes the soft and horny corals. Corals, like all anthozoans, have only a polyp form. There is no medusa stage in the life cycle.

The polyp of a stony coral resembles a sea anemone, although the individual polyps are generally smaller than those of anemones. The coral polyp sits in a **cup** on the surface of a calcareous exoskeleton secreted by the lower portion of the column and the basal disc. Extending inward from the wall of the cylindrical cup are several **calcareous septa,** which extend from the sides and base of the cup into folds of the basal tissue of the polyp. These tissue-covered septa partially subdivide the gastrovascular cavity inside the polyp into several chambers.

Most stony corals are colonial, and adjacent polyps are connected by lateral extensions of the body wall which cover the intervening stony skeleton of the colony. These sheets of lateral tissue also contribute to the formation of the skeleton by their secretions and serve to connect the gastrovascular cavities of adjacent polyps.

Many growth forms occur among the stony corals, and most species exhibit characteristic skeletons. A few species are solitary and occur as large individual polyps, like *Fungia* from the Pacific Ocean.

Some stony corals contain **symbiotic algae** (zooanthellae), and many interesting studies have been conducted on the physiological and biochemical interactions of these symbionts. Among the stony corals are several

reef-building species which are largely responsible for the formation of many coral reefs in warmer parts of the oceans, including those in the Bahama Islands, off the Florida Keys, and of the Great Barrier Reef of Australia.

Some examples of stony corals are *Fungia,* a solitary coral; *Astrangia danae,* the Atlantic star coral; *Oculina,* the eyed coral; and *Diploria* and *Meandrina,* brain corals.

The octocorals exhibit strong **eight-part** (octamerous) **radial symmetry** and have an **endoskeleton** consisting of separate microscopic pieces (spicules). A tough, horny organic material is also present in some species. This group is especially prominent in tropical waters and includes the sea fans, sea whips, the sea pens, the sea pansies, the organ pipe coral, and the precious red coral *(Corallium)* used for jewelry.

■ Study the demonstration materials illustrating several types of coral and draw several different types in figure 7.17. Identify each type of coral that you draw.

Demonstrations
1. Living sea anemones and/or corals in a marine aquarium.
2. Representative preserved anemones.
3. Assortment of preserved corals and dried coral skeletons.

Fig. 7.17 *Student drawing of corals.*

110   *Chapter 7*

Key Terms

Alternation of generations   the alternation of the sessile polyp and free-swimming medusa generations typical of the life cycle of the cnidarians.

Cnidocytes   specialized cells of cnidarians that produce and contain the nematocysts.

Coenosarc   the living portion of the tubular connecting portions of colonial cnidarians like *Obelia*. Consists of a simple cylinder of an outer epidermal tissue layer, an inner gastrodermal tissue layer, and an intermediate mesoglea surrounding a central gastrovascular cavity.

Dioecious   condition of an animal with sex organs borne in different individuals.

Diploblastic construction   the two-layered construction typical of the cnidarians. Consists of an outer epidermal tissue layer and an inner gastrodermal layer with an intervening noncellular mesoglea layer.

Epidermis   outer tissue layer protecting the surface of an animal from its environment.

Gastrodermis   inner tissue layer of animals bordering the digestive cavity.

Gastrovascular cavity   a central cavity of an animal that serves both for digestion and circulation, with a single mouth opening that serves both for entrance and exit. A type of incomplete digestive system.

Gonangium   a type of reproductive individual in colonial Hydrozoa, such as *Obelia;* produces free-swimming medusae.

Hydranth   feeding individual in a colonial hydrozoan, as in *Obelia*.

Medusa stage   free-swimming stage in the life cycle of many cnidarians. Usually bears gonads and produces gametes.

Mesoglea   gelatinous layer intermediate between the epidermal and gastrodermal layers of cnidarians, simple and noncellular in the Hydrozoa, containing cells and/ or fibers in the Scyphozoa and Anthozoa.

Monoecious   condition of bearing both sex organs in one individual.

Nematocysts   the stinging capsules produced by the cnidocytes of cnidarians. Characteristic of the phylum.

Nerve net   the diffuse, interconnected network of nerve cells in the cnidarians. Lacks ganglia or other nervous centers. A very primitive type of nervous system.

Perisarc   the nonliving outer covering secreted by the coenosarc of colonial hydrozoans. Surrounds the interconnecting coenosarc.

Planula larva   a simple, ciliated, sausage-shaped larval form produced by many cnidarians. Develops from the zygote or fertilized egg.

Polyp stage   the sessile (and usually asexual) stage in the life cycle of many cnidarians.

Radial symmetry   a body plan in which all body parts are arranged symmetrically around a central axis.

Scyphistoma   the inconspicuous polyp stage in the life cycle of certain scyphozoan medusae.

8

Platyhelminthes

Objectives

After completing the laboratory work in this chapter, you should be able to perform the following tasks:

1. Briefly outline the characteristics of the Phylum Platyhelminthes and identify some major advances in organization over the Phylum Cnidaria.

2. List and briefly characterize each of the three classes of the Phylum Platyhelminthes.

3. Describe the behavior of a free-living flatworm (planarian) such as *Dugesia* or *Planaria* and relate this behavior to the function of its sense organs.

4. Discuss the structure of the epidermis of a planarian and explain the location and function of cilia, rhabdites, and gland cells.

5. Identify the major structures that can be seen in microscopic cross sections at various levels of the body of a planarian; for example, anterior, pharyngeal region, and posterior.

6. Explain the organization of the reproductive system of a planarian and identify its principal reproductive organs.

7. Discuss the general morphology of the trematode *Clonorchis* and compare it with a free-living turbellarian such as *Planaria* or *Dugesia*.

8. Describe the life cycle of *Clonorchis* and identify the principal stages in microscopic preparations.

9. Describe the general morphology of the liver fluke *Fasciola* and identify its principal organs.

10. Identify in microscope slides the scolex, proglottids, and strobila of a tapeworm and explain the basic organization of a tapeworm.

11. Describe the life cycle of a tapeworm such as *Taenia* or *Dipylidium,* and identify its principal stages in microscopic preparations.

Introduction

The Platyhelminthes, or flatworms, are soft, wormlike animals with flattened, elongated bodies. They exhibit several important structural advances over the cnidarians, including **three distinct tissue layers** (triploblastic construction), **bilateral symmetry,** and several well-developed **organ systems.** Approximately 13,000 species of flatworms have been described.

The body parts of bilaterally symmetrical animals are arranged symmetrically along a mid-sagittal plane, so that the left and right sides are approximately mirror images of each other (see figure 9.1 in the next chapter). This type of symmetry is characteristic of most higher metazoans, except for adult echinoderms. Triploblastic body construction with **endoderm, mesoderm,** and **ectoderm** tissues is also characteristic of higher metazoans from Platyhelminthes to Chordata.

Flatworms lack the large central body cavity found in most higher animals. Instead, the interior of flatworms is typically filled with loosely packed **parenchyma** tissue with irregular spaces between the cells and clumps of cells. Since flatworms have no circulatory system or heart, body fluids percolate through these irregular interior spaces to bring nutrients and oxygen to the cells and to remove wastes from them. The body fluids are moved in part by muscular contractions. This type of organization without a central body cavity is called **acoelomate construction.**

Free-living flatworms (figure 8.1) have well-developed digestive, excretory, reproductive, nervous, and muscular systems, but, as noted previously, they have no circulatory system nor central body cavity. Some flatworms exhibit spiral cleavage and determinate development that suggest some affinities with the molluscs, annelids, and arthropods, animals that make up the group known as the protostomes.

Most flatworms, however, have become parasites and show many adaptations for parasitism, which include reduction and/or modification of some of their organ systems, and thus obscure clues to the possible evolutionary relationships of the phylum.

Fig. 8.1 *Free-living flatworms. (Courtesy of Carolina Biological Supply Company, Burlington, N.C.)*

Classification

The phylum includes three distinct classes:

Class Turbellaria (Free-living Flatworms)

Mainly free-living flatworms with bodies flattened dorsoventrally and a ciliated epidermis. Mouth usually ventral, leading into gastrovascular cavity; anus lacking. Freshwater and marine forms. Examples: *Dugesia, Planaria, Stenostomum, Leptoplana* (marine polyclad flatworm).

Class Trematoda (Flukes)

Parasitic animals with external **tegument** (cuticle) secreted by underlying cells. Ovoid body with one or two suckers for attachment to host; incomplete digestive system, usually with two main branches. With complex life cycle, involving larval stages and alternate hosts. Examples: *Clonorchis* (human liver fluke), *Fasciola* (sheep liver fluke), *Schistosoma*.

Class Cestoda (Tapeworms)

Elongate body with specialized scolex with hooks and/or suckers for attachment to host; body divided transversely into a series of proglottids; thick external tegument (cuticle); mouth and digestive tract absent. Usually with complex life cycle involving alternate hosts. Examples: *Dipylidium caninum* (dog tapeworm), *Taenia pisiformis* (dog and cat tapeworm), *Dibothriocephalus* (fish tapeworm).

Materials List
Living specimens
Dugesia or *Planaria*
Stenostomum
Cercariae
Preserved specimens
Whole tapeworms
Prepared microscope slides
Planaria, whole mount, representative cross sections
Stenostomum, whole mount
Clonorchis, whole mount
Fasciola, whole mount
Dipylidium, whole mount
Taenia pisiformis, scolex and representative sections
Taenia solium, scolex and representative sections (Demonstration)
Clonorchis, miracidium, sporocyst, redia, cercaria, metacercaria (Demonstrations)
Dibothriocephalus latus, scolex and representative sections (Demonstration)
Onchosphere (six-hooked) larva (Demonstration)
Cysticercus larva (Demonstration)
Plastic mounts
Dipylidium, whole mount (Demonstration)
Taenia pisiformis, representative sections (Demonstration)
Miscellaneous
Methyl cellulose solution
Beef liver

Free-living Flatworms: Class Turbellaria

A Planarian: *Dugesia*

Class Trematoda

Freshwater flatworms are often found on the underside of rocks, leaves, and sticks submerged in lakes, ponds, and streams. Although they are usually referred to as "planarians," the most common American freshwater flatworms are members of the genus *Dugesia* rather than the genera *Planaria* or *Euplanaria*.

Behavior and External Anatomy

■ Observe a living specimen and note its general size and shape. Is the worm uniformly pigmented, or does it have a distinctive pattern of pigmentation? Locate the **head,** the **eyes,** and the **auricles** (lateral projections of the head).

The eyes are light receptors but are not capable of forming real images. The auricles are well equipped with touch and chemical receptors. The head region also contains a concentration of nerve ganglia, which have some function in the processing of sensory information and serve as a primitive "brain."

Observe the smooth, gliding locomotion of the worm. This form of locomotion is due to the action of cilia on the ventral surface of the body, coordinated with rhythmic muscular contractions of the body. Note the behavior of the head and the auricles during locomotion. How is this behavior related to the function of sensory and nervous structures of the head? Touch the head of the worm gently with a clean dissecting needle or a toothpick. How does the worm react? Touch other body regions in a similar way and compare the reaction. Turn the worm over on its back. How does it react? Can you relate your observations of the worm's behavior to the structure of the head and the concentration of sense organs and nervous elements there? Can you make any conclusions about the possible advantages to an animal of having an anterior head with a concentration of sense organs and nervous tissue?

■ Place a small piece of fresh beef or pork liver in a dish containing one or more planarians. Observe their feeding behavior. Note the ventral **mouth** and the extension of the protrusible **pharynx** through the mouth when food is located. How does the worm locate the food? What sensory structures may be involved?

Internal Anatomy

■ Obtain a microscope slide with a stained whole mount of a planarian to study the anatomy further (figure 8.2).

Review the structures previously noted in the living specimen and also observe the **three-branched gastrovascular cavity.** Note the one anterior and two posterior branches of the cavity and the many smaller lateral branches, or diverticula.

■ Study also a microscope slide with cross sections through the anterior, middle, and posterior portions of the body (figure 8.3). In the section from the anterior portion of the body, locate (1) the **epidermis,** the external layer of cells surrounding the body; (2) the large, vacuolated cells of the **gastrodermis** lining the digestive tract; (3) the layers of **longitudinal** and **circular muscles** lying just inside the epidermis; (4) the large, irregularly shaped cells, the **parenchyma** tissue, which fills most of the interior space of the body; and (5) the two large **ventral nerve cords.**

Careful observation will reveal that the ventral epidermis is ciliated but the dorsal epithelium lacks cilia. Mucus and other types of **gland cells** are present in the epidermis and in the underlying mesenchyme. Many of the epidermal cells contain densely staining **rhabdites.** Rhabdites are small, rodlike bodies whose function is not yet fully understood. There is some evidence that they are discharged if the worm is attacked and swell up to form a slimy coat for defense. Some of the gland cells have long ducts that extend to the surface. These gland cells produce mucus for lubrication and other sticky materials for adhesion, capturing prey, and other functions. Among the parenchyma tissue, you should also find several small branches of the gastrovascular cavity surrounded by large vacuolated gastrodermal cells.

In the middle section through the buccal cavity of the planarian, find the large muscular **pharynx** lying within the **buccal cavity.** The pharynx has powerful muscles which allow it to be extended from the buccal cavity through the **ventral mouth** into the host's body to suck up fluids and soft tissues. Within the pharynx, locate several types of muscles: inner and outer layers of **circular muscle,** inner and outer layers of **longitudinal muscle,** and bands of **radial muscle** that extend across the pharynx from the outside to the central lumen. Find the **ciliated epithelium** on the outside of the pharynx. Many **gland cells** that produce proteolytic enzymes and mucus can also be found within the pharynx.

Study the section through the posterior portion of the body and note the several branches of the **gastrovascular cavity,** the ventral **nerve cords,** the circular and longitudinal **muscles,** and the **epidermis.**

Reproduction and Excretion

The reproductive system of *Dugesia* and other freshwater flatworms is small and difficult to observe except in special microscopic preparations. The reproductive organs are shown in figure 8.2, but most of them probably will not be visible in your slide. Since each worm has both male and female sex organs, planaria are **monoecious.**

Although planarians are monoecious, they are not normally self-fertilizing. In sexual reproduction, sperm is transferred from the male system of one worm by the male copulatory organ, the **penis,** to the **seminal receptacle** of the partner. The sperm subsequently moves to the oviduct of the female reproductive system where fertilization occurs. The fertilized eggs are later deposited outside the body in cocoons where they develop directly into young worms (figure 8.4).

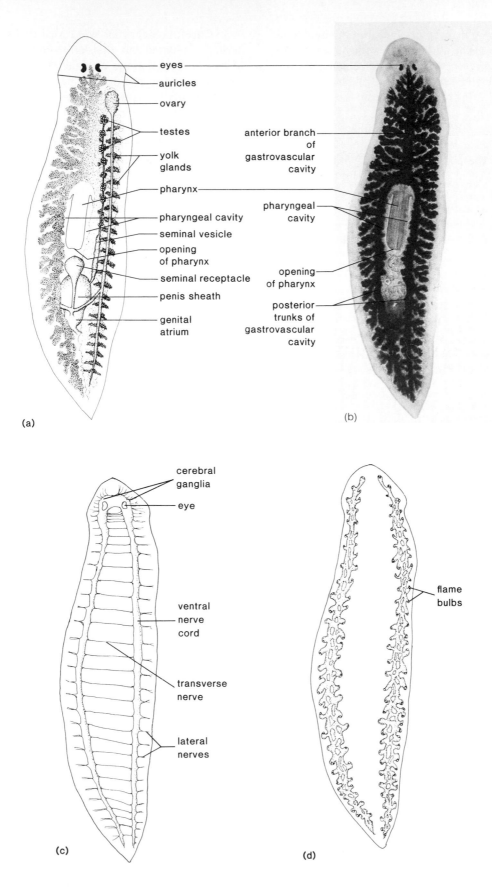

eyes

auricles

ovary

testes

yolk glands

pharynx

anterior branch of gastrovascular cavity

pharyngeal cavity

seminal vesicle

opening of pharynx

pharyngeal cavity

seminal receptacle

penis sheath

opening of pharynx

genital atrium

posterior trunks of gastrovascular cavity

(a)

(b)

cerebral ganglia

eye

ventral nerve cord

transverse nerve

lateral nerves

flame bulbs

(c)

(d)

Fig. 8.2   Planaria. *(a) Reproductive system. (b) Microscopic whole mount. (c) Nervous system. (d) Excretory system. (a. Photo courtesy of Carolina Biological Supply Company, Burlington, N.C.)*

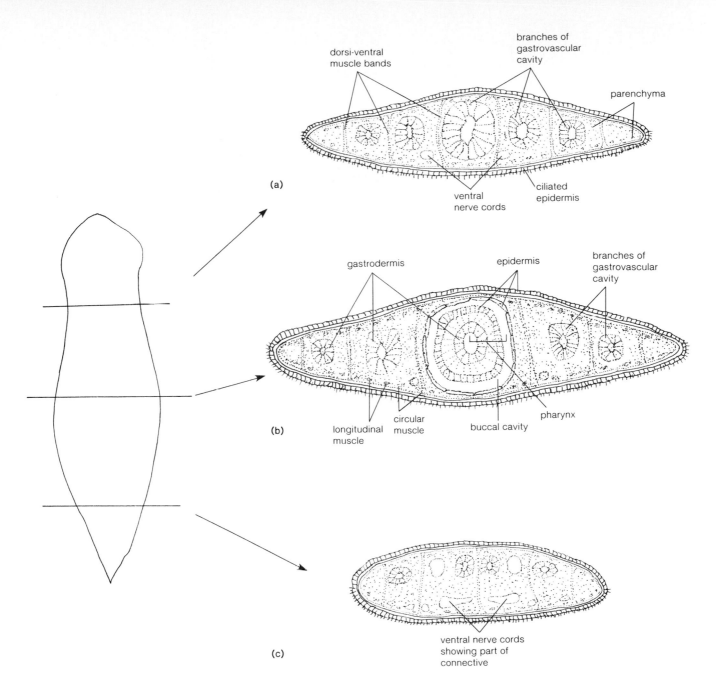

Fig. 8.3 Planaria. *Cross sections through three body regions: (a) anterior region, (b) pharyngeal region, and (c) posterior region.*

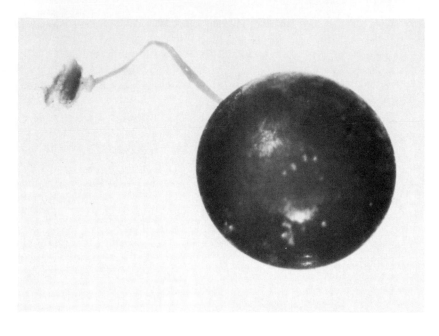

Fig. 8.4 Planaria. *Cocoon. (Courtesy of Carolina Biological Supply Company, Burlington, N.C.)*

In many species of planarians, however, the most common form of reproduction is **asexual.** A worm separates into two parts, and each part regenerates the missing structures. Planarians have great powers of regeneration, and even relatively small parts of a worm can develop into a complete animal. If time and materials permit, your instructor may be able to help you set up an experiment with regeneration in planarians.

The excretory/osmoregulatory system of *Dugesia* and other planarians is primitive. It consists of a system of **flame bulbs** (a type of protonephridia) interconnected by a system of collecting ducts leading to a posterior excretory pore. The excretory structures are difficult to observe, except in special microscopic preparations. The flame bulbs appear to function mainly in **osmoregulation.** The body fluids and cellular contents are hypertonic to the environment (contain more dissolved salts, etc.); thus, a planarian must constantly eliminate excess water from the body. Nitrogenous wastes, resulting from the breakdown of proteins and other nitrogenous matter in the food, are excreted, mainly as ammonia (NH_3), directly from the body cells.

Feeding and Digestion

Planarians, such as *Dugesia,* are chiefly carnivores and typically feed on protozoans and small animals, such as rotifers and small crustaceans. Food is ingested by the protrusible pharynx, which is extended through the midventral mouth. Do not confuse the opening of the muscular pharynx, which is extended through the mouth in feeding, with the actual midventral mouth opening. Proteolytic enzymes, secreted from glands near the tip of the pharynx, aid in penetration of a prey organism, such as a crustacean. The contents of the prey (a *Daphnia,* for example) can then be sucked into the muscular pharynx and passed into the gastrovascular cavity. The digestive system of a planarian is a gastrovascular cavity with a single opening that serves as both the entrance for food and the exit for waste materials.

■ How does this basic organization of the digestive system compare with that of a cnidarian or a higher animal, such as an earthworm or a frog? Which type of system would be more efficient? Why?

Digestion in a planarian is both **extracellular** (outside of the digestive cells) and **intracellular** (inside of the digestive cells). Digestive enzymes are secreted by **gland cells** in the gastrodermis that lines the gastrovascular cavity to assist in the breakdown of food materials. Later, small bits of food are engulfed by **phagocytic cells** in the gastrovascular lining. How many cell types can you identify in the gastrodermis in your cross section?

Muscular System

Other structures you should identify in the cross sections include the **longitudinal** and **circular muscle layers** just beneath the epidermis. Which of these layers lies closer to the epidermis? How can you relate these muscle layers to the locomotion that you observed in the living worms? Contraction of which layer would increase body length? How would this be helpful in locomotion?

Locate also in the cross sections the **dorsiventral muscle bands.** What is their function? Find the loosely packed parenchyma cells that fill most of the interior spaces. How are interior cells nourished? How are wastes removed from them? Near the ventral epidermis find the **two ventral nerve cords.**

The Flukes: Class Trematoda

The Human Liver Fluke: *Clonorchis (Opisthorchis) sinensis*

Class Trematoda

Members of the Class Trematoda are all parasitic and have well-developed **suckers** for attachment—one located in the region of the mouth, and one located on the ventral surface. The outer covering of the trematode body is highly modified and lacks cilia. The outer layer, the **tegument** (formerly called the cuticle), is a syncytial extension of underlying cells embedded in the body wall. Electron microscope studies have revealed that the tegument has a complex structure. Tapeworms (Class Cestoda) also have a tegument with a similar structure.

The tegument is a nonciliated syncytial tissue that serves an active role both in protecting trematodes from the digestive enzymes of the hosts and in the uptake of nutrients from the host gut. The tegument is an excellent example of morphological and physiological adaptation of a parasite for its very special mode of life.

Clonorchis sinensis (figure 8.5), sometimes also called *Opisthorchis sinensis,* is a common and important human parasite in certain parts of the world, particularly in the Orient. Like many other trematodes, this species has a complex life cycle involving several hosts and a series of larval stages (see figure 8.6).

■ Obtain a prepared microscope slide with a stained whole mount of an adult fluke and observe its size, shape, and general morphology under your stereoscopic microscope. Observe the **oral sucker** surrounding the mouth at the anterior end.

Behind the mouth is a muscular **pharynx,** a short **esophagus,** and two **intestinal caeca.** Note that the digestive tract has a single opening, the **mouth,** and thus is an **incomplete digestive system.** A bilobed **cerebral ganglion** ("brain") lies on the dorsal side of the pharynx (small and difficult to see in most slides). Near the branching of the intestinal caeca, note the **posterior sucker.** Immediately behind the posterior sucker, along the midline of the body, is the long, coiled **uterus** containing many eggs.

Posterior to the uterus lie the many-branched **testes** where the sperm are produced. Along the lateral margin of the body in its midregion, observe the many small **yolk glands.** The yolk glands connect with the ovary by means of two delicate **yolk ducts.** The **ovary** is a single, small structure located near the center of the body. It is connected with the **seminal receptacle,** which serves to store sperm received during copulation.

■ Study figure 8.6 and the demonstration materials provided to illustrate the life cycle of *Clonorchis.*

Note that the life cycle includes parasitic stages in three different hosts: *human, snail,* and *fish.* In order to survive, a parasite with such a complex life cycle including several hosts must have some effective means of transfer from one host to the next. Unless the proper host is available at the appropriate time, the life cycle cannot continue, and the parasite will die. This principle is used as the basis for the control of many parasitic diseases of humans and animals, such as malaria and schistosomiasis.

The adult liver fluke lives in the bile duct of humans and of certain other carnivorous animals. The host in which the adult (sexually mature) stage of a parasite resides is designated as the **definitive host.** All other hosts in the life cycle are termed **intermediate hosts.**

Human infections of *Clonorchis* occur from eating raw fish. Adult worms live in the bile ducts of the liver, and fertilized eggs are released into the bile duct. The eggs pass into the small intestine and are later voided in the feces of the host. If the feces get into water, the eggs may be eaten by certain species of snails (**first intermediate host**). Inside the digestive system of the snail, the egg hatches into a larval form called a **miracidium.** The miracidium lives in the tissues of the snail, passing through several other larval stages (**sporocyst, redia,** and **cercaria**) and reproducing asexually to produce thousands of new larvae. The last larval stage, the **cercaria,** escapes from the snail and swims in the water until it contacts the **second**

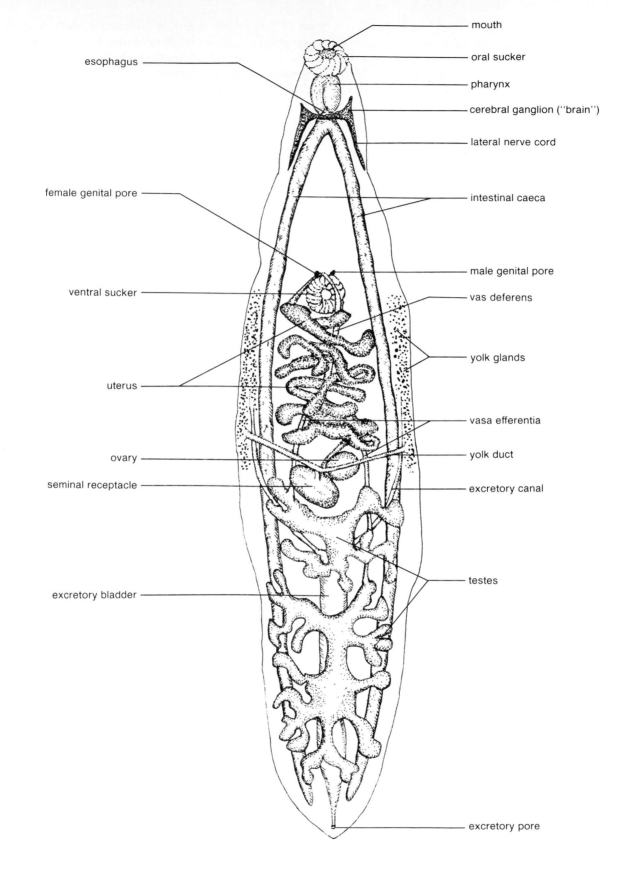

mouth

oral sucker

esophagus

pharynx

cerebral ganglion ("brain")

lateral nerve cord

female genital pore

intestinal caeca

male genital pore

ventral sucker

vas deferens

yolk glands

uterus

vasa efferentia

ovary

yolk duct

seminal receptacle

excretory canal

excretory bladder

testes

excretory pore

Fig. 8.5 Clonorchis, *whole mount.*

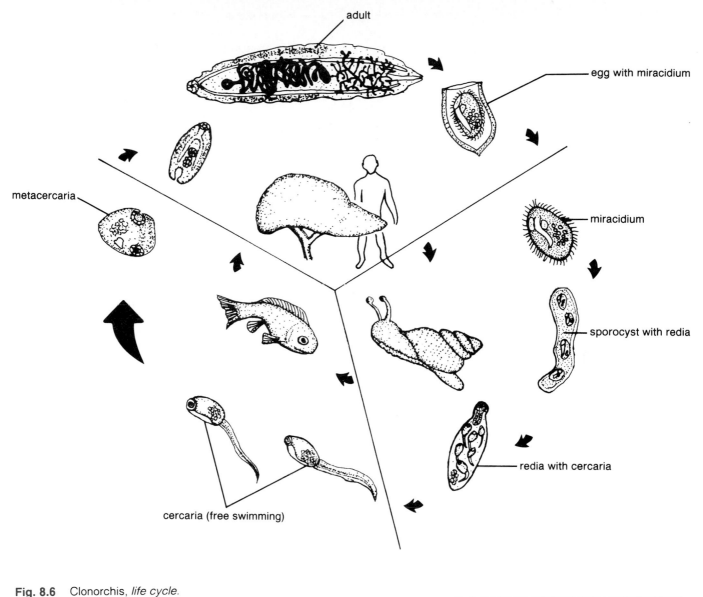

adult

egg with miracidium

metacercaria

miracidium

sporocyst with redia

redia with cercaria

cercaria (free swimming)

Fig. 8.6 Clonorchis, *life cycle.*

intermediate host, certain species of fish. When the fish is contacted, the cercaria burrow through the skin, shed their tails, and encyst to form still another stage, the **metacercaria.** If raw or improperly cooked fish containing metacercaria is eaten by a human or another appropriate **definitive host,** the cyst walls are digested, and the metacercaria are released. Subsequently, they migrate into the bile ducts of the liver where they develop into adult flukes to complete the life cycle.

The Sheep Liver Fluke: *Fasciola hepatica*
Class Trematoda

This large trematode is generally similar in structure to *Clonorchis*, although it is considerably larger in size, and its reproductive system is slightly more complex (figure

8.7). *Fasciola hepatica* is a common parasite of sheep and cattle, but it also occasionally parasitizes other mammals, including humans. The adult flukes live mainly in the bile ducts and cause the breakdown of the adjacent liver tissue, producing the disease called "liver rot."

The life cycle of *Fasciola hepatica* is of historic importance since it was the first to be worked out for a digenetic (two-host) trematode (figure 8.8). Adult worms live in the bile duct of a sheep and produce many **eggs** that pass from the bile duct to the intestine and are deposited with the feces. In water or warm, moist conditions, the **miracidium larva** develops within the egg and escapes. This microscopic ciliated larva must burrow into the appropriate species of snail within a few hours to continue its development. Unless it finds an appropriate host, the miracidium dies.

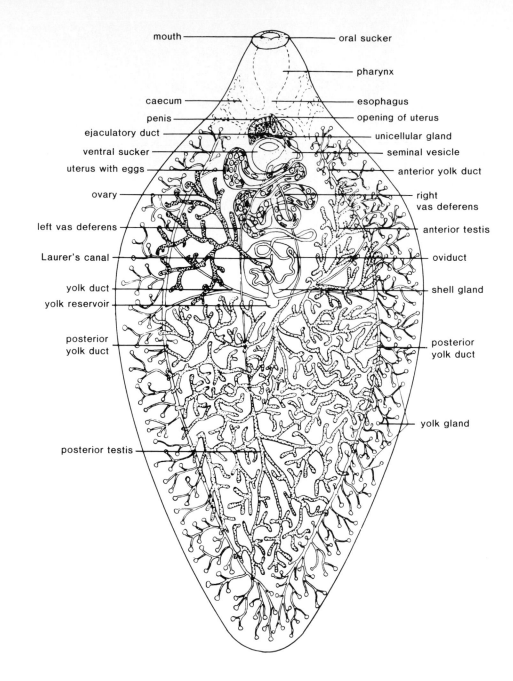

Fig. 8.7 Fasciola, *whole mount.*

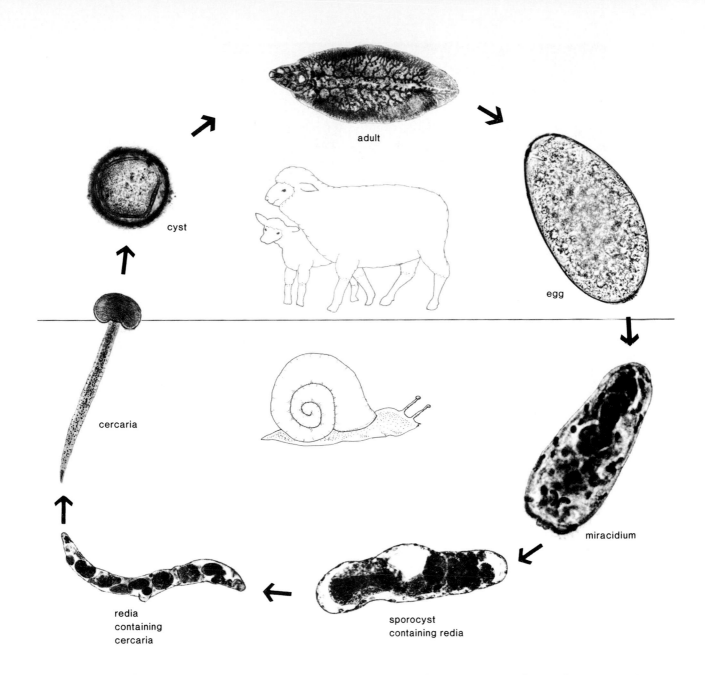

adult

cyst

egg

cercaria

miracidium

redia
containing
cercaria

sporocyst
containing redia

Fig. 8.8 Fasciola. *Sheep liver fluke. Life cycle stages. (Photos courtesy of Carolina Biological Supply Company, Burlington, N.C.)*

mouth
anterior sucker

penis extending
from genital
pore

ventral sucker

Fig. 8.9 Fasciola. *Ventral surface of anterior end. Note sculpturing on tegument. Magnification 28×. (Scanning electron micrograph by Louis de Vos.)*

In the appropriate snail, which serves as the **intermediate host,** the miracidium develops into a **sporocyst.** Within the sporocyst several germ cells develop into the next larval stage, the **redia** (plural: rediae). The redia also reproduces asexually and produce still more rediae. The last generation of rediae produces another larval stage, the **cercaria** (plural: cercariae). The cercariae escape from the snail and become free-swimming. The cercariae swim until they reach some aquatic plants, settle on the plant, lose their tails, and encyst as **metacercariae.** The encysted larvae can survive for several weeks on the plants, and if the plant and metacercariae are eaten by a sheep or other appropriate final or **definitive host,** the cyst wall is digested and the larvae burrow through the intestinal wall to the body cavity and thence to the liver. During their migration through the liver, the young adults feed on the liver tissue.

■ Observe the demonstrations of the various stages in the life cycle and try to understand the sequence of stages with the aid of figure 8.8.

■ Obtain a slide with a stained whole mount of an adult *Fasciola* and note its general shape. How does its shape differ from that of *Clonorchis?* Locate the **anterior sucker** around the mouth and the nearby **ventral sucker.** Between the two suckers, find the **genital pore** (figure 8.9).

The digestive system includes the anterior **mouth,** a muscular **pharynx,** a short **esophagus,** and two branches of the **intestine,** each of which has many smaller lateral branches, the **intestinal caeca** (singular: caecum). What would you expect to be the main purpose of the intestinal caeca? Both male and female reproductive systems are present as in *Clonorchis.* With the aid of figure 8.7, locate the principal organs of each system.

The female organs (mainly found in the anterior half of the body) include: **ovary, yolk glands, yolk ducts, yolk reservoir, shell gland, oviduct, uterus** with eggs, and the **opening of the uterus** just inside the genital pore.

The male organs include: **testes** (one anterior and one posterior), **vas deferens** (how many?), **seminal vesicle** (how many?), **ejaculatory duct,** and a muscular **penis.**

Demonstrations
1. Slides with adult stages of some other representative trematodes.
2. Slides representing the stages in the life cycle of *Clonorchis* and/or *Fasciola.*
3. Living cercaria.

The Tapeworms: Class Cestoda

The Dog Tapeworm: *Dipylidium caninum*
Class Cestoda

Tapeworms are highly specialized internal parasites and show several important adaptations for their parasitic mode of life. The adult worms inhabit the intestines of various species of vertebrate animals, and the larvae live in the tissues of some alternate host. In general, the life cycles of tapeworms, or cestodes, are less complicated than those of the trematodes.

Fig. 8.10 Taenia taeniaeformis, *from cat. Magnification 300✕. (Scanning electron micrograph by Fred. H. Whitaker.)*

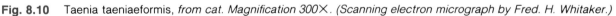

The flat, ribbonlike body of a tapeworm is typically divided into many sections called **proglottids,** and the body is divided into three major regions: an anterior **scolex,** a specialized holdfast organ with suckers and/or hooks (figure 8.10); followed by a narrow **neck,** which contains the budding zone where, at the posterior end, new proglottids are produced asexually; and the **strobila,** the rest of the long body, usually consisting of many proglottids. Although the tapeworms superficially appear to be segmented, the proglottids are not generally believed to represent true body segments because of the way in which they are formed and because each proglottid is a complete reproductive unit within itself. Many zoologists therefore view the body of a tapeworm as comparable to a colony, or a chain of individuals, rather than as being actually segmented.

■ Obtain a prepared microscope slide or a plastic mount with the scolex and some representative proglottids of the dog tapeworm, *Dipylidium caninum* (figures 8.11 and 8.12). Locate and study the following structures: the **scolex,** the **neck,** and the **strobila.** On the scolex, find the **rostellum,** the raised tip of the scolex, which bears several rows of **hooks,** and four lateral **suckers.** The hooks and suckers aid in attachment to the intestinal wall of the host intestine. Note the absence of a **mouth, pharynx,** and **digestive system.**

Where would you find the youngest proglottids in an intact tapeworm? The oldest? Study several proglottids in different stages of development and note the different degrees of elaboration of the reproductive systems. Identify **immature, mature,** and **ripe proglottids** (see figure 8.11).

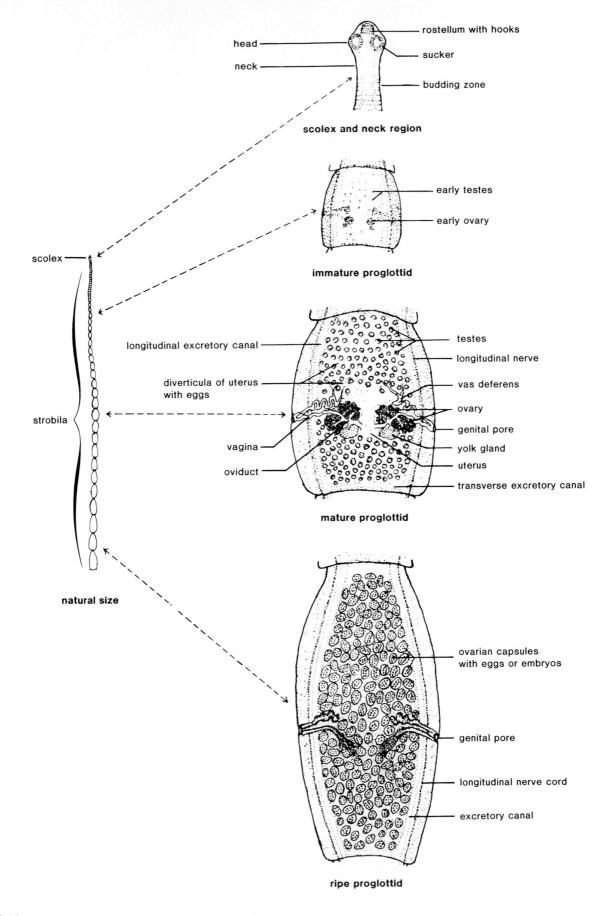

Fig. 8.11   Dipylidium, *representative sections of body.*

Chapter 8

Fig. 8.12 Dipylidium caninum. *(a) scolex, (b) mature proglottid, and (c) ripe proglottid. (Courtesy of Carolina Biological Supply Company, Burlington, N.C.)*

Select a mature proglottid for more detailed study and identify the **genital pore, vagina, oviduct, yolk glands, uterus, testes, vasa deferentia** (singular: vas deferens), **excretory canals,** and **longitudinal nerve cords.** Note the complete absence of any digestive structures in the tapeworm. How do you suppose tapeworms obtain their nourishment?

Study also some of the ripe proglottids and observe the many ovarian capsules containing several eggs or embryos. Is the number of eggs per capsule always the same? Locate some of the transitional segments between the mature and the ripe proglottids to observe the progressive atrophy of certain of the reproductive organs.

The life cycle of *Dipylidium* includes two larval stages, a six-hooked **onchosphere larva** and a **cysticercoid larva** ("bladder worm"). Ripe proglottids containing eggs and embryos pass out of the host intestine in the feces.

The eggs are ingested by flea larvae and hatch into onchospheres inside the intestinal wall. Later they develop into cystercoid larvae as the fleas become adults. A dog or cat may be infected by nipping a flea. Children sometimes become infected with *Dipylidium,* presumably after being licked by a dog or by ingesting eggs deposited in the soil.

A Dog and Cat Tapeworm: *Taenia pisiformis*
Class Cestoda

Taenia pisiformis (figures 8.13 and 8.14) is another tapeworm commonly studied in zoology laboratories. Adult tapeworms of this species occur in the small intestine of dogs and cats. The larval stages are found in the liver and mesenteries of rabbits.

| scolex and anterior proglottids | immature proglottids | mature proglottids | ripe proglottids |

Fig. 8.13 Taenia pisiformis. *Representative sections. (Courtesy of Carolina Biological Supply Company, Burlington, N.C.)*

■ Obtain a microscope slide and/or plastic mount with a scolex and representative proglottids of *Taenia pisiformis*. Identify the **scolex, neck, hooks,** and **lateral suckers.** Observe that *Taenia pisiformis,* like other tapeworms, has no mouth or digestive system. How does this tapeworm obtain its nourishment?

Locate and identify **immature, mature,** and **ripe proglottids.** Select a mature proglottid on your slide and study its internal structure. Find the **uterus, yolk glands, oviduct, vagina, seminal receptacle,** and **genital pore** comprising the female reproductive system. Among the male reproductive structures, locate the **testes, vas efferens, vas deferens,** and the copulatory organ, the **cirrus.** Also find the lateral **excretory canals.** Two **longitudinal nerve cords** should be visible lateral to the excretory canals.

Adaptations for Parasitism

Tapeworms provide a good example of the adaptations of animals for a special mode of life. Tapeworms exhibit several structural and physiological features that increase

Fig. 8.14 Taenia pisiformis, *mature proglottid.*

their chances of survival as internal parasites of vertebrates. List six different adaptations related to their parasitic mode of life that you have been able to observe or to identify from your reading.

1. _____

2. _____

3. _____

4. _____

5. _____

6. _____

Demonstrations

1. Microscopic preparations of other tapeworms such as *Taenia solium* (pork tapeworm), *Taeniarhynchus saginatus* (beef tapeworm), and *Dibothriocephalus latus* (broad or fish tapeworm).
2. Onchosphere (six-hooked) larva.
3. Cysticercus larva.
4. Preserved whole tapeworms.

Key Terms

Bilateral symmetry　body form with parts arranged symmetrically along a mid-sagittal plane.

Cercaria　motile, free-swimming, tadpolelike larval stage in the life cycle of certain trematodes. Body resembles miniature fluke. Plural: cercariae.

Definitive host　final host in the life cycle of a parasite with more than one; host in which sexually mature parasite develops.

Intermediate host　host in the life cycle of a parasite which bears a larval or immature stage of that parasite.

Miracidium　ciliated larval stage in the life cycle of many trematodes. Develops from the fertilized egg and gives rise to the sporocyst stage.

Proglottid　reproductive body division of a tapeworm. Since each proglottid contains a complete set of male and female organs, these units are often considered analogous to a complete individual, and thus the tapeworm as a colony.

Redia　larval stage in the life cycle of certain trematodes. Formed by the sporocyst and produces many cercaria.

Scolex　specialized holdfast organ of the cestodes (tapeworms); with hooks and/or suckers for attachment. Plural: scoleces or scolices.

Sporocyst　saclike larval stage in life cycle of some trematodes. Develops from miracidium and produces several redia.

Strobila　the major portion of the body of a tapeworm excluding the scolex. Includes all of the proglottids.

Tegument　living syncytial protective outer covering of the cestode and trematode body; formerly called the cuticle; secreted by large underlying cells with numerous tubular cytoplasmic connections with the tegument.

Triploblastic construction　body type with three distinct tissue layers: endoderm, mesoderm, and ectoderm. Characteristic of flatworms and all higher metazoan phyla.

9

Hints for Dissection

Objectives

After completing the laboratory work in this chapter, you should be able to perform the following tasks:

1. Explain the objectives of dissection and the importance of good dissection techniques.
2. List the basic instruments needed for dissection, explain the use of each, and demonstrate their proper use and care.
3. Distinguish between the anatomical terms anterior and posterior; dorsal and ventral; transverse plane, frontal plane, and sagittal plane.

Introduction

Many of the animals to be studied in the following exercises will require dissection for the study of internal structures. To gain the most from these studies, it is important that you follow directions carefully and that you develop good dissection skills. Both attributes are essential to your training as a scientist. You must take special care in each dissection to avoid damaging important structures before you have had an opportunity to complete your study.

Some important anatomical terms are illustrated in figure 9.1 and are defined among the **Key Terms** at the end of this chapter. Some understanding of basic anatomy is essential for success in zoology, regardless of your future specialization. You should review figure 9.1 and the list of Key Terms frequently, to reinforce your understanding of anatomy and proper terminology, as you proceed through the remaining exercises in this book.

The chief objective in the dissection of a specimen is to expose body parts for the study of their structure and their relationships to other parts. Therefore, it is important to proceed carefully with each dissection and to avoid cutting or removing anything, unless so instructed in the printed directions or by your laboratory instructor. Read through the instructions completely before you begin each dissection. This will save you time later and will often prevent disappointing results.

Body parts should be parted carefully along natural lines of separation and attachment wherever possible. This can often be done best with forceps, probe and seeker, the handle of a scalpel, or with a finger, without cutting.

Good quality dissection instruments are essential to good laboratory work. Your dissection kit should include the following items:

> Scissors, fine points (or one fine and one blunt point), about 4–6 inches long
> Forceps, fine points, about 4–5 inches long
> Scalpel with replaceable blade
> Extra scalpel blades
> Two teasing needles
> Probe and seeker, about 6 inches long
> Plastic centimeter ruler
> Dropping pipette

Scissors should be first-quality chrome or stainless steel and have the blades joined by a screw rather than a rivet. Some common dissecting instruments are illustrated in figure 9.2.

Dissecting instruments should be kept clean and sharp at all times. A small oilstone and a piece of emery cloth or fine sandpaper should be used to keep cutting edges and needle points sharp. Two types of scalpel are in common use today. One type has a permanent blade that can be resharpened, and the second type has a separate handle and replaceable blades that are not usually sharpened. Blades on scalpels with replaceable blades should be replaced often to ensure a good edge. Sharp scalpels and scissors allow clean cuts and minimize damage to adjacent tissues. Always wash and fully dry your dissecting instruments with a paper towel after each use. With proper care, a good set of instruments will last for many years.

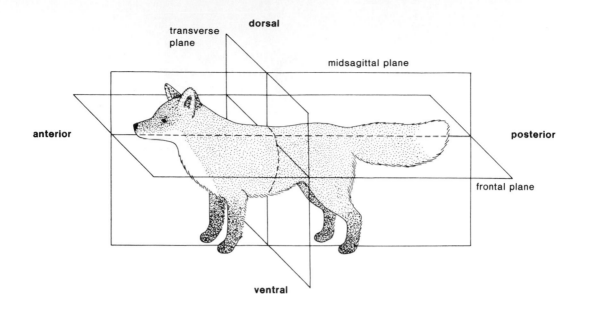

Fig. 9.1 *Basic anatomical directions and planes.*

Fig. 9.2 *Dissecting instruments.*

The scalpel is used for making incisions in the body wall and occasionally for sectioning interior structures. Hold the scalpel upright with the handle nearly perpendicular to the surface to be cut (figure 9.3) and make a clean forward incision while supporting surrounding tissues with your fingers. Be conscious of internal structures and avoid cutting too deeply. If you have to saw with the scalpel to cut the tissue, your blade is too dull.

Forceps and dissecting needles should be used for loosening, lifting, and moving various structures to facilitate study and to expose underlying parts. It is important to gain some idea of the thickness of the covering or surface layer to be cut and its relationship to other important structures nearby before starting your dissection. Consult any available charts, photographs, models, and demonstration dissections of the animal to be studied. A few

Fig. 9.3 *Proper use of the scalpel.*

minutes spent in such preliminary orientation often will prevent later disappointment. Also remember that your laboratory work will be evaluated by your instructor partly on the basis of your care and skill in dissecting.

Read the dissection instructions in the laboratory manual and follow them carefully. Always remember that **direct study** of internal anatomy gained from your dissections of the animals is the primary objective. The illustrations in this manual are intended to **help** you in your dissections and study but are **not intended to substitute** for the study of real specimens. Refer to the illustrations frequently, but focus your study primarily on the animal, not on the pictures.

In some of the exercises, you are also asked to make drawings of the animals you study and dissect. Take care to be accurate and neat in your drawings so that they will be helpful to you for later study and review. Remember that your laboratory notes and drawings serve a purpose similar to your lecture notes. They provide a record to help you to recall actual structures during your later study and reviews. They help you to retain and to recall mental pictures of the things actually seen through the microscope and during your dissections.

You should also use the pages provided for *Notes and Sketches* at the end of each chapter for any additional things that you observe or need to remember.

Key Terms

Aboral   away from or opposite to the mouth.

Anterior   toward the front or head end; the opposite of posterior.

Asymmetry   an irregular arrangement of body parts; without a central point, axis, or plane of symmetry.

Bilateral symmetry   an arrangement of body parts on opposite sides of a central plane (midsagittal plane), which divides the body into two symmetrical halves (mirror images).

Caudal   toward the tail or tail end; the opposite of cephalic.

Cephalic   of or pertaining to the head; the opposite of caudal.

Cranial   relating to the skull or cranium.

Cross section   sections of the body cut on any transverse plane; such sections are perpendicular to the sagittal and frontal planes.

Deep   pertaining to structures away from the surface of the body; the opposite of superficial.

Distal   away from the center or point of attachment; the opposite of medial.

Dorsal   relating to the back or upper surface; the opposite of ventral.

Frontal plane   plane parallel to the dorsal and ventral surfaces of the body, which bisects a bilaterally symmetrical animal into upper and lower halves.

Lateral   toward the side; the opposite of medial.

Longitudinal   lengthwise; parallel to the long axis.

Medial   toward the sagittal plane or center of the body; the opposite of lateral.

Median   located in or near the sagittal plane.

Oral   relating to the mouth.

Peripheral   toward the outer surface.

Posterior   the hind part (rear) of the body; the opposite of anterior.

Proximal   toward the center or point of attachment; the opposite of distal.

Radial symmetry   arrangement of body parts symmetrically around a central axis; any plane through the central axis divides the body into symmetrical halves (mirror images).

Sagittal plane   any longitudinal plane passing from the head to tail. The midsagittal plane bisects a bilateral animal into two symmetrical halves (mirror images).

Superficial   located near the surface of the body; the opposite of deep.

Transverse plane   any plane perpendicular to the sagittal and frontal planes. Sections of the body cut on a transverse plane are called cross sections.

Ventral   relating to the belly or underside; the opposite of dorsal.

10

Pseudocoelomate Animals

Objectives

After completing the laboratory work in this chapter, you should be able to perform the following tasks:

1. List the seven animal phyla commonly grouped together as pseudocoelomate animals and list five of the characteristics they have in common.

2. Describe the external morphology of the roundworm *Ascaris* and locate the chief external features in preserved specimens. Distinguish between mature male and female specimens.

3. Describe the male and female reproductive systems of *Ascaris* and identify the principal reproductive organs in dissected specimens.

4. Describe the anatomy of the rotifer *Philodina* and locate its principal organs in a specimen or drawing.

5. List and briefly explain the importance of five other pseudocoelomate animals.

Introduction

The roundworms, Phylum Nematoda (Nemathelminthes), are sometimes studied together with several other animal groups in a large heterogeneous Phylum Aschelminthes, but recent investigations have led most zoologists to consider these animal groups as separate phyla because of their many differences. The most prominent similarity among these groups is the pseudocoelom, a central body cavity lying between the endoderm and mesoderm layers, derived from the embryonic blastocoel. Other common features include a cylindrical body with an external cuticle, microscopic size, a complete digestive tract with a mouth and an anus, a specialized pharynx, a protonephridial excretory system, and a fixed number of cells (or nuclei) in the adult. The latter condition results from the cessation of mitoses in the adult, thus limiting the number of cells (or nuclei) present in each mature individual.

Numerous exceptions to these "common" features have been found and have led to the present uncertainty about the true relationships among these animal groups. For instance, recent electron microscopic investigations have shown that some of these groups have a much reduced body cavity, and, in some cases, this cavity may not actually be a pseudocoelom. Much more study is needed on these enigmatic animals.

Seven groups (or phyla) of animals are commonly recognized in this assemblage, including the Nematoda, Rotifera, Gastrotricha, Kinorhyncha, Nematomorpha, Acanthocephala, and the Entoprocta (see figure 10.1). An eighth phylum, the Gnathostomulida, is sometimes included, but many zoologists believe that the Gnathostomulids are actually more closely related to the flatworms (Phylum Platyhelminthes) than to the pseudocoelomates.

Classification

The seven phyla most commonly grouped among the pseudocoelomates are briefly described below.

Phylum Nematoda (Nemathelminthes or Roundworms)

Free-living and parasitic worms with an **elongate cylindrical body tapered at both ends,** a **triradiate pharynx,** and a **specialized excretory system.**

Phylum Rotifera (Rotifers)

Microscopic animals, found primarily in freshwater habitats, with an anterior ciliated locomotory and feeding organ, the **corona,** and a specialized internal grinding organ, the **mastax.**

Phylum Gastrotricha (Gastrotrichs)

Microscopic aquatic animals with an **external cuticle** covered by scales or spines, and **patches or tracts of cilia** on the epidermis of the ventral surface.

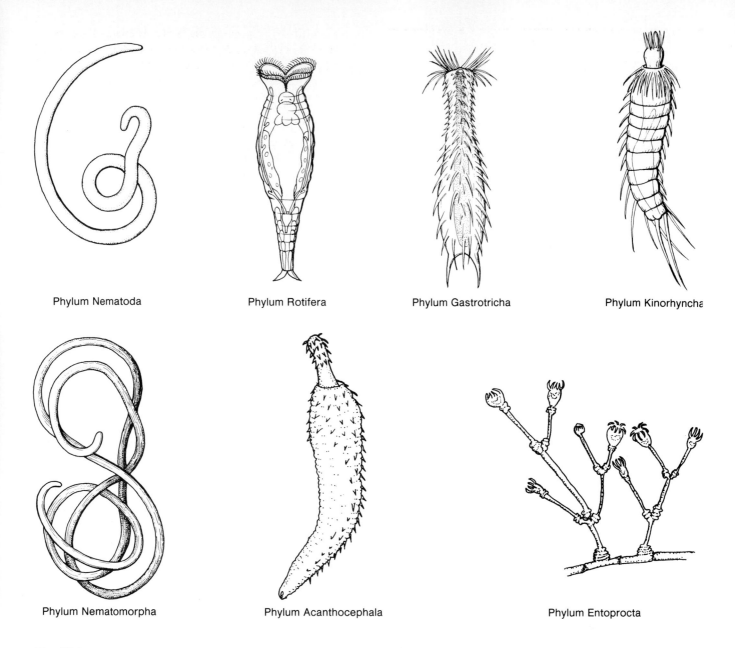

Phylum Nematoda

Phylum Rotifera

Phylum Gastrotricha

Phylum Kinorhyncha

Phylum Nematomorpha

Phylum Acanthocephala

Phylum Entoprocta

Fig. 10.1 *Chief phyla of pseudocoelomate animals.*

Phylum Kinorhyncha (Kinorhynchs)

Microscopic marine animals with strong **superficial segmentation** (segmented cuticle), a **retractile head with spines,** and **lateral spines** along the body.

Phylum Nematomorpha (Gordiacea or Horsehair Worms)

Slender, elongate worms with a **cylindrical body rounded at the ends,** a **reduced digestive tract,** and **larvae parasitic in insects** or other arthropods.

Phylum Acanthocephala (Spiny-headed Worms)

Parasitic worms with adults living in the intestines of vertebrates and larvae in arthropod hosts; **anterior proboscis with hooks** for attachment to gut wall of host; **no digestive tract.**

Phylum Entoprocta (Entoprocts)

Tiny, sessile animals with a **cup-shaped body** attached to the substrate by a **stalk,** solitary or colonial, with a **U-shaped digestive tract** and a **crown of ciliated tentacles** surrounding the mouth and the anus. Most species are marine.

Phylum Nematoda

The Nematoda and Rotifera are the largest, best known, and most important among these phyla. Some of the common representatives of these two phyla are included in this laboratory exercise to illustrate some of the important features of pseudocoelomate animals.

A Parasitic Roundworm: *Ascaris lumbricoides*

External Anatomy

Ascaris lumbricoides (figure 10.2) is a large roundworm parasite of humans and swine. The adult worms range in length from about 15–40 cm. Living specimens are yellow or pink in color, but preserved specimens usually range from pink to brown.

■ Study a preserved specimen to determine some of the characteristic features of roundworms, members of the Phylum Nematoda.

Observe the elongate cylindrical body that tapers at both ends. The anterior end is more pointed than the posterior, which tends to be slightly blunt. The triangular **mouth**, surrounded by three distinct **lips**, is located at the anterior end of the body (see figure 10.3). Under your dissecting microscope observe the three lips and the mouth. Also locate the **anus**, which appears as a transverse slit on the ventral side of the body near the posterior end. The ventral location of the anus is a convenient means of distinguishing the dorsal and ventral surfaces of the worm. You should be able to find four longitudinal lines extending lengthwise along the body wall: one dorsal, one ventral, and two lateral in position. These longitudinal lines are more apparent in living specimens than in preserved specimens, and their significance will be better understood after your study of the microscopic anatomy of the worm later in this exercise.

A further examination of the external features of your specimen should reveal its sex. Most species of nematodes are **dioecious** (sexes separate), and the males and females usually exhibit certain external morphological differences. Male specimens of *Ascaris lumbricoides* typically exhibit a curved posterior end and a pair of **copulatory spicules** extending from the anus (see figure 10.2). Female worms have a ventral **genital pore** about one-third the distance from the anterior end. From your observation of the location of the copulatory spicules in the male, what can you conclude about the location of the male genital opening? Examine several specimens of *Ascaris lumbricoides* to determine whether you can find any other morphological differences between the sexes.

Internal Anatomy

■ After you have completed your study of the external features of *Ascaris*, carefully dissect the worm by making a longitudinal slit down its dorsal side. Take care with your incision to avoid damaging the internal organs. Pin down the body wall on each side of the worm in a wax-bottom dissecting pan to expose fully the internal structures, and cover your specimen with a little water to keep the internal organs moist and flexible.

Caution: Although the *Ascaris* specimens you will receive in the laboratory have been preserved in formalin, the eggs found in the female specimens are enclosed in a very resistant "shell." They may not all be killed, even after several weeks in formalin. Although acquiring an infection in this way is a remote possibility, you should always follow good laboratory safety practices and avoid putting your hands in your mouth after handling live or preserved animals, and you should wash your hands after you complete your work.

Observe the long, ribbonlike **intestine** extending most of the length of the body and the numerous tangled threadlike coils surrounding the middle part of the intestine (see figure 10.2). The female reproductive system is Y-shaped, and each branch of the Y consists of a long, threadlike **ovary** connected with a larger **oviduct** and a still larger **uterus.** The distinction among these three organs is not apparent externally but is easy to observe in microscopic cross sections. The two uteri join to form a single midventral muscular tube, the **vagina,** which opens to the exterior via the female genital pore, or **vulva,** observed previously.

The male reproductive system is not branched, as in the female reproductive system, and it consists of a single, continuous tubular structure, including a threadlike **testis** connected with a **vas deferens,** or sperm duct, larger in diameter, a still-larger-in-diameter **seminal vesicle,** and a short, muscular **ejaculatory duct,** which empties into the **cloaca** at the posterior end of the digestive tract.

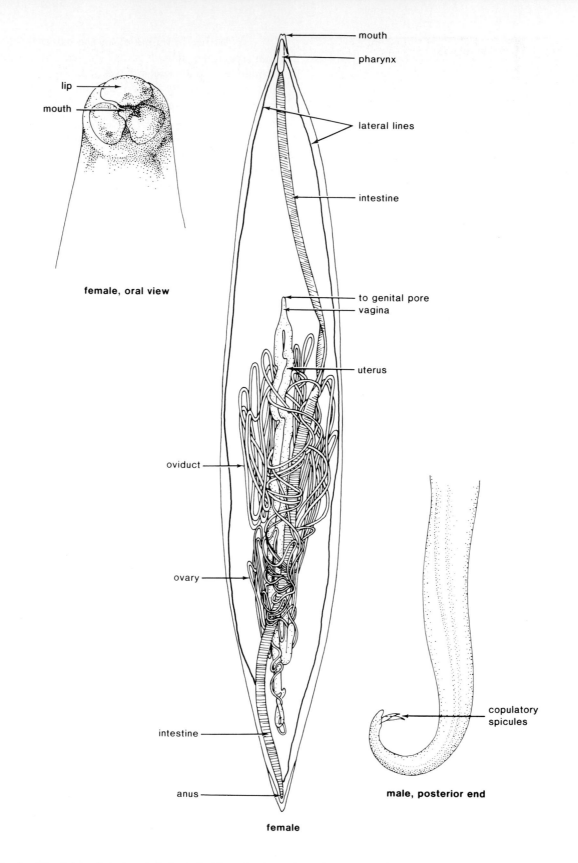

female, oral view

mouth
pharynx
lateral lines
intestine
to genital pore
vagina
uterus
oviduct
ovary
intestine
anus

female

lip
mouth

copulatory
spicules

male, posterior end

Fig. 10.2 Ascaris, *internal anatomy of dissected female plus selected external features of male and female.*

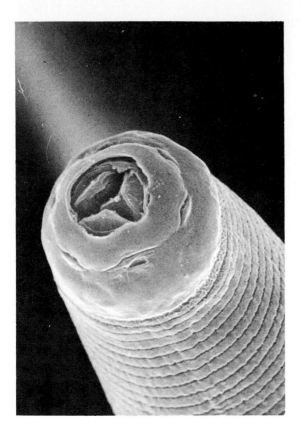

Fig. 10.3 Nematode. *Mouth, lips, and anterior end showing triangular mouth surrounded by three teeth as in* Ascaris. *Magnification 5,200×. (Scanning electron micrograph by Louis de Vos.)*

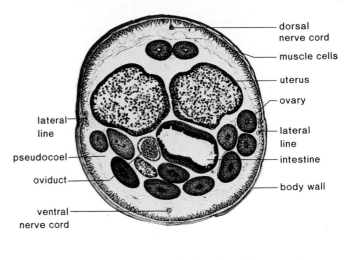

Fig. 10.4 Ascaris, *cross section, female. (Courtesy of Carolina Biological Supply Company, Burlington, N.C.)*

The digestive system consists of the ribbonlike **intestine,** which is attached anteriorly to a short muscular **pharynx** into which the mouth opens, and posteriorly to the **cloaca.**

Microscopic Study

Prepared microscope slides with cross sections through male and female worms illustrate further aspects of nematode structure (figure 10.4).

Examine a cross section of a female worm and identify the thick **cuticle** covering the body and the underlying layer of **hypodermis,** which secretes the cuticle. Observe the many **longitudinal muscle bands;** there are no circular muscles present in roundworms.

This anatomical feature produces the peculiar whiplike movements often exhibited by living roundworms. Free-living nematodes are sometimes called "whipworms." Also associated with the body wall are the **dorsal** and **ventral nerve cords** and the two **lateral lines.** Locate within each of the lateral lines the **excretory canal,** part of the peculiar specialized excretory system of the nematodes.

Nematodes lack protonephridia and possess a unique system of gland cells connected to a canal system that empties through a ventral excretory pore. This system, called the **renette gland,** is located ventrally near the pharynx. The function of this unique excretory system is not well understood, but *Ascaris* is known to excrete nitrogenous waste in the form both of ammonia and urea.

Observe the thin-walled intestine and note that the intestinal wall consists of a single layer of large **gastrodermal cells.** The body cavity of *Ascaris* is a **pseudocoelom,** since it lacks a mesodermal layer around the gut.

In the cross section of the worm, the female reproductive system should be represented by numerous round or oval structures of various sizes. The **ovaries** are small in diameter and bear some resemblance to tiny wagon wheels. A layer of columnar ovarian cells surrounds a central core, or **rachis.** The **oviducts** are slightly larger in diameter than the ovaries and have a central opening, or lumen, for the passage of eggs. If the section was cut through the region of the vagina, you could observe a single, large median structure, usually filled with eggs. Sections posterior to the vagina region usually show two large uteri, also filled with eggs.

Cross sections of male specimens (figure 10.5) exhibit features similar to female specimens except for the reproductive structures. The reproductive system is usually represented by several sections through the **testes,** which are small round structures containing many spermatogonia. Several sections through the larger **vas deferens** should also be present in the section. You should find

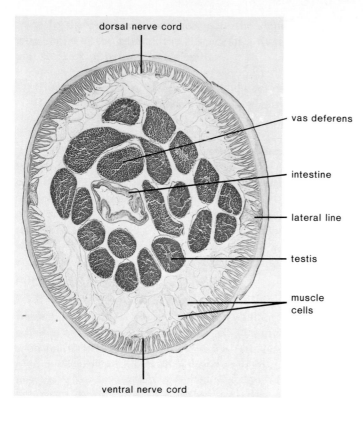

Fig. 10.5 Ascaris, *cross section, male. (Courtesy of Carolina Biological Supply Company, Burlington, N.C.)*

Fig. 10.6 Ascaris. *Eggs with developing larval worms. (Courtesy of Carolina Biological Supply Company, Burlington, N.C.)*

numerous spermatocytes in the vas deferens of mature male specimens. A large **seminal vesicle** containing mature spermatozoa should also be present.

Life Cycle

Ascaris lumbricoides has a direct life cycle with no intermediate hosts. Mature female worms in the intestine of the host produce eggs that are expelled in the feces. The fertilized eggs develop into larvae in warm, moist soil. If these eggs containing the larvae (figure 10.6) are ingested by an appropriate host, development may continue. Hatching occurs in the small intestine and the larvae emerge, burrow through the intestinal wall, and are carried to the liver of the host via the hepatic portal vein. From the liver the larvae migrate to the lungs. Later they move up through the bronchi and trachea and reach the oral cavity. From there the worms are swallowed and are carried to the small intestine. Sexual maturity is reached within about two months, when fertilized eggs begin to be released. Fertilization occurs after copulation between male and female worms in the host's intestine.

A Parasitic Roundworm: *Trichinella spiralis*

Another common parasitic nematode is *Trichinella spiralis*, the trichina worm. *Trichinella* is the causative agent of the disease **trichinosis** and occurs commonly in humans, swine, bears, rats, and other mammals. The disease is usually contracted by humans through eating raw or inadequately cooked pork or bear meat.

■ Observe a microscope slide showing encysted worms in the skeletal muscles of a pig or a rat (figure 10.7).

Encysted worms in infected meat may be released through digestion of the cyst walls by enzymes in the human digestive tract. The juvenile worms mature sexually in about two to three days and then mate. The females burrow into the wall of the small intestine and shortly begin to produce living offspring. A single female may produce up to 1,500 young in her three-month life span. These juvenile worms enter the host's lymphatic system and are distributed through the body via the circulatory system. The young worms burrow mainly in the active skeletal muscles, where they grow to about 1mm in length and become enclosed in cysts. Although such encysted worms cannot continue to develop unless the infected tissue is eaten by another suitable host, the encysted worms may remain alive within the cysts for several years.

Hookworms: *Ancylostoma duodenale* and *Necator americanus*

Ancylostoma duodenale and *Necator americanus* are the two most common and most serious human hookworms. *A. duodenale* is predominantly a northern species common in Europe, North Africa, northern China, and Japan, while *N. americanus* is primarily a tropical species occurring in many warmer areas of the world. Until recent years *N. americanus* was a very serious public health menace in

muscle tissue

Trichinella larva

Fig. 10.7 Trichinella. *Larval worm, encysted in muscle tissue. (Courtesy of Carolina Biological Supply Company, Burlington, N.C.)*

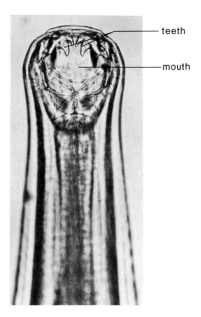

teeth

mouth

Fig. 10.8 Ancylostoma, *mouth and buccal cavity showing teeth. (Courtesy of Carolina Biological Supply Company, Burlington, N.C.)*

the southeastern United States. Improved sanitation and control measures in the past few decades, however, have greatly reduced the severity of the hookworm problem.

Both species are principally human parasites, but are sometimes also found in other animal species. Both species also exhibit specialized **teeth** and/or **cutting plates** around the mouth (see figure 10.8), which aid in the attachment of the adult worms to the wall of the small intestine. The body of *N. americanus* is smaller and more slender than that of *A. duodenale*.

■ Study the demonstration slides of hookworms.

Life Cycle

The life cycles of the two species are basically similar. The adults live in the small intestine, where they attach to the intestinal mucosa and feed on blood and tissue fluids. Adult females release large numbers of eggs each day (5,000–20,000). The eggs are released in the host feces, and the embryonated eggs develop and hatch in the dung. The young, free-living larvae feed on bacteria and other debris, grow, molt several times, and reach the infective stage in about five days. During this time, they live in the upper layer of soil and frequently migrate to the surface.

They are attracted to heat and are stimulated by physical contact. These attributes facilitate contact and penetration of exposed human skin—like a bare foot.

The larvae burrow through the bare skin until a blood or lymph vessel is reached. In the bloodstream they are carried to the heart and then to the lungs. There they burrow into the alveoli. From the lungs they are passed up the bronchi and trachea to the throat. From the throat they may be swallowed and pass through the digestive tract to the small intestine, where they undergo a final molt and attach to the mucosa. Sexual maturity is reached in about six weeks.

Pinworm: *Enterobius vermicularis*

The most common nematode parasite of humans in the United States is the pinworm *Enterobius vermicularis* (figure 10.9). One epidemiological study showed that as many as one-third of American preschool children may be infected.

■ Obtain a microscope slide with a stained whole mount of an adult *Enterobius* and observe its size, shape, and internal organs.

mouth

Fig. 10.9 Enterobius. *Female. (Courtesy of Carolina Biological Supply Company, Burlington, N.C.)*

Adult female pinworms are 8–13 mm long and have a pointed tail. They have enormous reproductive potential; a single female can lay more than 16,000 eggs in her lifetime.

Male and female adult worms live in the host intestine where they attach to the walls. Gravid females migrate posteriorly to the anus from which they often emerge at night to deposit their eggs on the surrounding skin. Infections and reinfections usually arise from ingestion of eggs transferred from contaminated bedding or from fingernails contaminated from scratching the anal area.

A Free-living Roundworm: *Anguillula aceti*

The vinegar eel (figure 10.10), *Anguillula aceti (Turbatrix aceti),* is a tiny, free-living nematode sometimes occurring abundantly in vinegar. It was more common in the past than it is today, since most commercial vinegar is now pasteurized and also has preservatives added to prevent the growth of vinegar eels, and other undesirable flora and fauna. The worms are most abundant in the bottom sediments of unpasteurized vinegar and other fermented fruit juices. Vinegar eels thrive in such acid conditions and feed on the yeast and bacteria growing in the sediment. This combination of yeast and bacteria is often called the "mother" of the vinegar.

■ Examine a sample from the culture of the worms and observe their active, whiplike swimming movements. How can you relate their mode of locomotion to what you have learned about the anatomy of the nematodes? What kind of muscles are involved?

Select a few large worms for further study and place them in a small drop of vinegar on a clean microscope slide. If the worms are too active for study, they may be killed by briefly warming the slide over an incandescent lamp or a small flame. The specimens may then be observed unstained, or they may be lightly stained with hematoxylin or Nile blue sulphate in 70% alcohol.

Observe the blunt anterior end with its terminal mouth and the pointed posterior end. What is the location of the anus? Identify the straight digestive tract consisting of the **mouth, pharynx, pharyngeal bulb** containing a three-part pharyngeal valve, an **intestine,** a thin-walled **rectum,** and a subterminal **anus.** In male specimens, the rectum opens into a **cloaca,** which in turn opens through the anus.

Select a large female worm for the study of the reproductive system. Identify the **female genital pore** on the ventral side of the body. This genital opening serves both for receiving sperm during copulation and for the release of the offspring, which are born live in this species. Extending posteriorly from the genital opening is a blind sac, the **seminal receptacle,** believed to serve for the temporary storage of sperm after copulation. Anterior to the genital pore is a single large **uterus,** which may contain several developing young. At its anterior end (about one-third the distance from the anterior end of the body), the uterus is attached to a smaller **oviduct,** which bends posteriorly and connects with the threadlike **ovary,** where the eggs are produced.

Males may be recognized by their **copulatory spicules** and by the different structure of their reproductive system. Observe the opening of the **genital pore** into the **cloaca,** the small spherical spermatozoa in the **vas deferens,** and the filamentous **testis,** which may extend almost to the middle third of the body, where it terminates in a slight enlargement. Observe a demonstration of a male worm that has been treated with strong formalin that causes the copulatory spicules to project from the anus.

Other Free-living Roundworms

Many species of free-living roundworms live in soil, decaying vegetation, and in many aquatic habitats, such as lakes, ponds, and streams. An easy way to find aquatic roundworms is to collect small bits of decaying plants from the edge of a pond or stream. Place the plant material in a small finger bowl and gently tease apart the material with dissecting needles and/or forceps. You should be able to identify the nematodes under a stereoscopic microscope by observing their shape and characteristic whiplike movements. A stereoscopic microscope equipped with a substage lamp to provide transmitted light through the sample works best.

You may also wish to examine some small samples of soil rich in organic material, "black dirt," in a similar manner for soil nematodes. Pieces of a raw or boiled potato

142

Fig. 10.10 Anguillula aceti, *the vinegar eel.*

Lateral view of female

mouth
pharyngeal bulb
intestine
endodermal lining
pharynx
pseudocoel
developing embryos
uterus
egg
oviduct
genital pore
ovary
seminal receptacle
intestine
rectum
anus

Lateral view of male

intestine
copulatory spicule
testis
cloaca

Pseudocoelomate Animals

143

may be used for bait. Place the potato pieces in a moist place in a garden, under a board or rock, or beneath a log in the woods for a few days. You can then tease apart the potato bait to search for nematodes immediately or incubate the pieces for a few days in a sterile petri dish or test tube in a warm place. The latter procedure will enable any eggs and/or larvae to develop.

Phylum Rotifera

A Rotifer: *Philodina*

Rotifers are common and abundant freshwater animals. They are important constituents of the plankton of lakes, ponds, and streams, and are significant as a food source for many species of fish and other filter-feeding animals.

Philodina (figure 10.11) is a common North American rotifer that illustrates many of the typical features of the Phylum Rotifera.

■ Obtain a living specimen of *Philodina* and prepare a wet mount on a microscope slide to study its behavior and appearance.

Living specimens are much superior to fixed and stained specimens for the study of general anatomy of rotifers because of the serious contraction and distortion almost always encountered during their fixation and staining.

Observe the elongate cylindrical body, which can be divided into three general regions—the **head**, the **trunk**, and a posterior **foot.** Superficially, the body is divided into several segments, usually about 16. The rotifers have no true segmentation, but the cuticle covering the body is often divided into a number of superficial segments. Observe the telescoping of the segments when the animal retracts its head. Note also the large ciliated corona at the anterior end. Most rotifers have a conspicuous **corona,** which serves both for locomotion and for feeding. In *Philodina,* the corona consists of two large ciliated **trochal discs** and a posterior band of cilia, the **cingulum.** Special **retractor muscles** serve to retract the corona. Observe also, on the head, the dorsal fingerlike process, the **rostrum.** The rostrum is believed to be a sensory structure. Observe the long, tapering **foot** with two **spurs** and four retractable **toes** at the posterior end of the body.

The body wall is covered externally by a thin **cuticle** secreted by a syncytial **hypodermis.** Cell membranes are generally lacking in the organs and tissues of adult rotifers, and specific organs typically have a fixed number of nuclei. Although there are no definite muscle layers associated with the body wall of rotifers, several distinct **muscle bands** connect various parts of the body. As in members of related phyla, the central body cavity of rotifers is a **pseudocoelom.**

The **mouth** is a funnel-shaped opening located at the base of the corona. It receives food particles swept down by ciliary currents created by the corona. Connecting with the mouth is a muscular **pharynx,** which encloses a conspicuous grinding apparatus, the **mastax.** The mastax is a distinctive feature of the rotifers and can be easily identified in a living rotifer because of its active, almost constant movement. In *Philodina,* the mastax is specialized for grinding plankton, periphyton, and detritus, but other types of mastax are found in different species of rotifers, and their structure appears to be closely related to the food habits of the various species.

Posterior to the pharynx is a short **esophagus** surrounded by the large **salivary glands.** The esophagus opens into a large, thick-walled **stomach** where digested food is absorbed by the gastrodermal lining. Behind the stomach is a short **intestine** leading into the **cloaca.** The cloaca empties through the **anus.**

Most of the reproductive system of *Philodina* can be seen only in special preparations, but a pair of large, yolk-forming glands, the **vitellaria,** are readily visible in living specimens. Observe the two large vitellaria in the posterior part of the trunk region of your specimen. How many transparent nuclei can you identify? The vitellaria are syncytial like most other rotifer organs and have a fixed number of nuclei. Other parts of the reproductive system include two small anterior **ovaries** (difficult to see), which connect via small ducts to the large vitellaria, and two tiny oviducts that lead from the vitellaria to the cloaca.

Males are unknown in the group of rotifers to which *Philodina* belongs (Order Bdelloidea), and the eggs produced develop exclusively by parthenogenesis. Certain other types of rotifers, however, do exhibit normal sexual reproduction.

Other Pseudocoelomates

Phylum Nematoda

Rhabditis maupasi. A small nematode found in the soil and often as a parasite in earthworms.

Wuchereria bancrofti, filaria worm. Important human parasite in many warm parts of the world. Causative agent of the disease elephantiasis, which is characterized by a massive accumulation of fluids after the blockage of lymphatic circulation, causing great enlargement of the limbs or other parts of the body.

Heterodera schactii, the sugar beet nematode. A specialized plant-eating nematode that feeds on the root of the sugar beet.

Phylum Rotifera

Epiphanes senta. A common, relatively large rotifer often cited in textbooks as a representative rotifer. Exhibits a single median ovary and vitellarium; males are small and simplified in structure.

Asplanchna. This genus includes several species of planktonic rotifers with a saclike body, an incomplete digestive tract, and a well-developed corona.

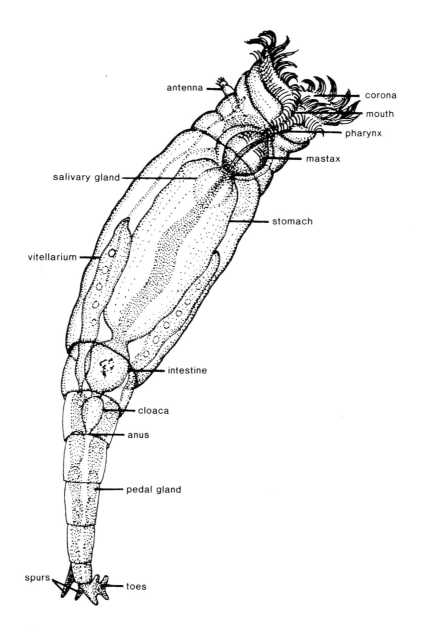

Fig. 10.11 Philodina, *a rotifer.*

Phylum Nematomorpha

Paragordius. Widely occurring genus of horsehair worms. Adult worms found in marshes, ponds, streams, and lakes, often in tangled masses of several worms. Larvae occur as internal parasites of beetles, grasshoppers, and crickets.

Demonstration
Specimens of representative pseudocoelomates.

Key Terms

Cloaca   a chamber at the posterior end of the digestive tract in certain types of animals that serves to receive wastes from the digestive tract, gametes from the reproductive tract, and/or wastes from the excretory system.

Corona   a specialized organ at the anterior end of rotifers, usually ciliated, which often serves for locomotion and in food gathering. The circular motion of the coronal cilia in many species of rotifers provides the basis for the common name, "wheel animals," often given to rotifers.

Cuticle   a thick, noncellular protective covering secreted by an underlying layer of epithelial cells, the hypodermis.

Hypodermis   a type of tissue made up of epidermal cells that secrete an overlying cuticle.

Mastax   a specialized grinding organ found in the pharyngeal region of rotifers.

Pseudocoelom   a type of internal body cavity that lacks a mesodermal lining around the digestive tract. It arises embryonically as a remnant of the blastocoel.

Trichinosis   a parasitic disease of humans and certain other carnivorous mammals caused by the nematode parasite *Trichinella spiralis.* Usually acquired by eating rare or uncooked pork or bear meat.

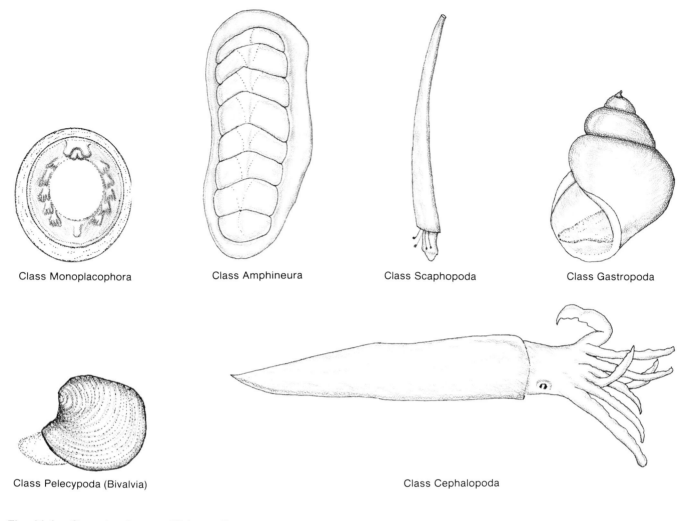

Class Monoplacophora	Class Amphineura	Class Scaphopoda	Class Gastropoda

Class Pelecypoda (Bivalvia)

Class Cephalopoda

Fig. 11.1 *Six major classes of living molluscs.*

Class Monoplacophora

Primitive molluscs with a conical, one-piece shell and definite evidence of segmentation in several organ systems. Known only from fossil forms and from a few recent specimens dredged from great depths in the sea at a few locations during the past few years. All marine. Example: *Neopilina*.

Class Amphineura

The chitons, or mail shells. Body oval shaped, with a large, flat foot and having eight dorsal calcareous plates. Algal feeders, mainly in the marine intertidal zone. Examples: *Chaetopleura, Chiton, Cryptochiton.*

Class Aplacophora

The solenogasters. A small group of peculiar wormlike marine molluscs with no head or shell, and with a reduced foot; with an anterior mouth and a posterior anus. Some recent authorities suggest splitting the Aplacophora into two classes, the Solenogastres and the Caudofoveata. Examples: *Neomenia, Crystallophrisson (Chaetoderma).*

Class Scaphopoda

The toothshells. Body elongate dorsoventrally and encased in a tapered, tubular, one-piece shell open at both ends. Burrowing marine molluscs found in sandy and muddy sea bottoms. Scaphopod shells were once used by certain Native Americans on the American Pacific coast as money. Example: *Dentalium.*

Class Gastropoda

The snails, slugs, whelks, and limpets. Animals with a long, flat foot; a distinct head with eyes and tentacles; and a dorsal visceral mass usually housed in a spiral shell. The asymmetrical growth of the visceral mass and the overlying mantle is responsible for the spiral shape of the shell. Gastropods comprise the largest and most successful class of molluscs. Ecologically, they are the most versatile molluscs with freshwater, marine, and terrestrial species. Some gastropods are carnivores (feed on animal tissues), some are herbivores (feed on plant material), and still others are parasites. Examples: *Helix* (European garden snail), *Littorina* (periwinkle), *Busycon* (whelk), *Physa,* and *Lymnaea* (freshwater snails).

Class Pelecypoda

The clams, mussels, oysters, and scallops. Body laterally compressed, small foot, no head, body contained in a bivalve (two-piece) shell hinged on the dorsal side. Most pelecypods are adapted for a sedentary life in marine or freshwater habitats. Typically they are filter-feeders and have specialized gills that serve to trap suspended food particles. Examples: *Anadonta* and *Unio* (freshwater mussels), *Mercenaria* (hardshell clam), *Crassostrea* (an oyster), *Pecten,* and *Aquipecten* (scallops).

Class Cephalopoda

The squids, cuttlefish, nautili, and octopi. Cephalopods are the most advanced molluscs and possess a large head with conspicuous eyes; mouth surrounded by eight or ten or more fleshy arms or tentacles; elongate body; shell often internal, reduced, or absent. They are typically active marine animals, preying on various fish, molluscs, arthropods, and worms. Examples: *Loligo* (squid), *Octopus, Nautilus* (chambered nautilus).

Materials List

Living specimens
    Physa or similar freshwater snail
    Representative molluscs (Demonstration)
Preserved specimens
    Freshwater mussel
    Loligo
        Loligo, liquid mount of dissection
    Helix
        Helix, plastic or liquid mount of dissection
    Dextral and sinestral gastropod shells (Demonstration)
    Dried pen of squid (Demonstration)
    Nautilus shell (Demonstration)
    Representative molluscs (Demonstration)
Prepared microscope slides
    Mussel shell, cross section (Demonstration)
    Glochidium larva (Demonstration)
    Radula of gastropod (Demonstration)
Chemicals
    Carmine or carbon powder

Demonstration

Living and preserved representatives of various classes of molluscs.

The Freshwater Mussel

Class Pelecypoda

Freshwater mussels are sedentary animals that live in or on the bottoms of lakes, ponds, and streams. The particular species available for laboratory study vary from time to time and from place to place, and may include any of several large species in the genera *Anodonta, Lampsilis, Elliptio,* and *Quadrula.* The basic anatomy of the hard-shell marine clam *Mercenaria* (formerly called *Venus*) is also very similar to the following description and can be used instead of the freshwater mussel if desired.

■ Obtain a mussel in a dissecting pan and note the hard **bivalve shell.**

The two valves are joined along the dorsal surfaces by an elastic **hinge ligament.** The exterior surface of the valves is covered by a dark, horny material, the **periostracum.** Observe the concentric **lines of growth** on the exterior surface of the shell, which are formed as the mantle secretes new material at the edge of the shell.

The shell consists of three layers, the exterior **periostracum,** a middle **prismatic layer,** and an inner **nacreous layer.** The exterior periostracum is made up of a structural protein, conchiolin, which retards dissolution of the shell by the slightly acidic waters in which freshwater mussels are typically found. The middle prismatic layer consists of crystalline calcium carbonate ($CaCO_3$) and provides strength. The inner nacreous layer ("mother-of-pearl") consists of numerous layers of $CaCO_3$ and is iridescent. Near the anterior end of each valve is a raised portion, the umbo, which represents the oldest part of the valve. At the edge of the shell, between the valves, you should be able to observe the **incurrent siphon** (lower) and the **excurrent siphon** (upper), two openings between the edges of the mantle (see figure 11.2).

In figure 11.2, locate the two large muscles that hold the valves together, the **anterior** and **posterior adductor muscles.** They work antagonistically to the hinge ligament, a strong chitinous structure on the dorsal edge of the valves that tends to keep them open. You must cut through the anterior and posterior adductor muscles (carefully) to gain access to the interior organs.

■ Check to see whether your mussel has a wooden peg inserted to keep the valves spread apart. If not, it will be necessary for you to pry open the valves by inserting the handle of your forceps or the handle of your scalpel between the valves along the ventral edge. (Be careful of the blade!) Twist the handle, and when the valves are sufficiently separated, place a wedge between them so that they are separated about one-fourth to one-half inch. Carefully insert your scalpel, blade first, into the space between the left valve and the closely applied mantle in the region just below the anterior adductor muscle (see figure 11.2). Keep the blade close against the shell, loosen the mantle from the valve, and cut the large anterior adductor muscle. Repeat the procedure at the posterior end and cut the posterior adductor muscle. Now carefully lift the loosened left valve, separating the mantle from it as you lift. **Warning:** The heart is located in the pericardial cavity near the dorsal side of the shell; take care to avoid damage to this region.

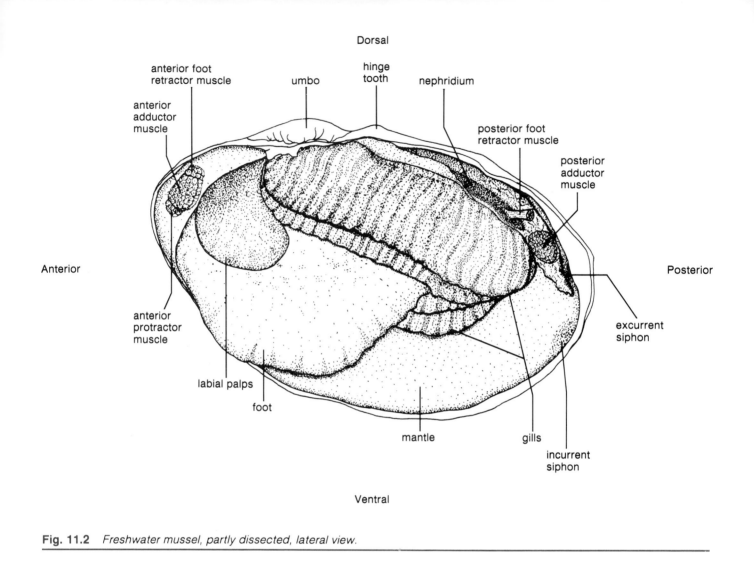

Dorsal

anterior foot
retractor muscle

anterior
adductor
muscle

umbo

hinge
tooth

nephridium

posterior foot
retractor muscle

posterior
adductor
muscle

Anterior

Posterior

anterior
protractor
muscle

labial palps

foot

mantle

gills

excurrent
siphon

incurrent
siphon

Ventral

Fig. 11.2 *Freshwater mussel, partly dissected, lateral view.*

Feeding, Digestion, and Respiration

The thin layer of tissue that lines each valve is the **mantle.**
It attaches to the inside of each valve along the **pallial line,**
which is located about half of the distance from the edge
of the valve to the center. Enclosed within the mantle is a
space, the **mantle cavity,** which contains the other organs.
Along the edge of the mantle identify the ventral **incurrent siphon,** with **sensory papillae** lining its borders, and
the dorsal **excurrent siphon.** Locate the muscular **foot,** the
gills, the **labial palps,** and the **mouth** located at the base
of the palps (figure 11.3).

The large **gills** play an important role in both respiration and feeding. Each gill consists of a double fold of
tissue suspended in the mantle cavity (see figure 11.4).
Each gill fold is a **lamella.** The lamellae of each gill (left
and right) join ventrally to form a **food groove** and dorsally to form a **suprabranchial chamber,** which carries
water posteriorly to the excurrent siphon. The gills contain blood vessels and obtain some oxygen from the incoming water currents. The mantle is also vascularized
and serves as a respiratory organ.

Cilia on the surface of the mantle and gills create
water currents in the mantle cavity and draw in water
through the incurrent siphon. Food particles contained in
the incoming water are filtered from the water as it passes
through the gills and are trapped in mucus secreted by
glands in the gill tissue. The entrapped food particles are
collected in the food tubes of the gills and transported anteriorly by highly coordinated ciliary movements to the
labial palps and into the mouth. Along this route, nonfood
particles are sorted out and eliminated.

■ If live mussels or clams are available in the laboratory, you can observe the coordinated ciliary movements on the surface of the gills by placing a few
particles of carbon powder (or carmine powder) on
the moistened surface of a gill. Make certain that the
surface of the gills is moist (not wet) before you apply
the carbon and take care to apply only a few particles
of carbon as a marker. When working with live mussels or clams, be sure to add water to your specimen
frequently to keep the soft internal tissues moist and
flexible. The soft tissues dry out rapidly when they
are exposed to the air.

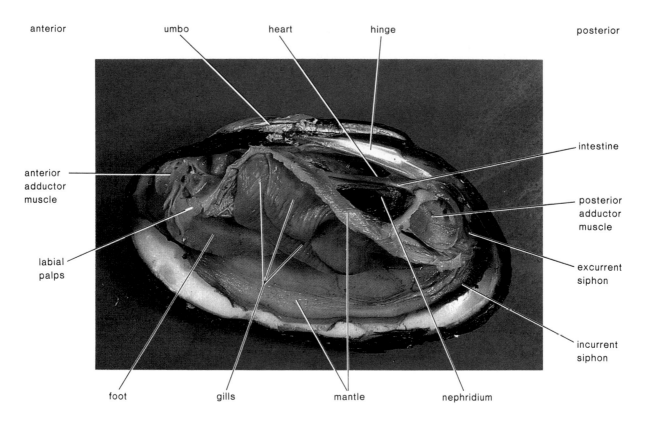

Fig. 11.3 *Freshwater mussel, part of mantle removed. (Photograph by Ken Taylor.)*

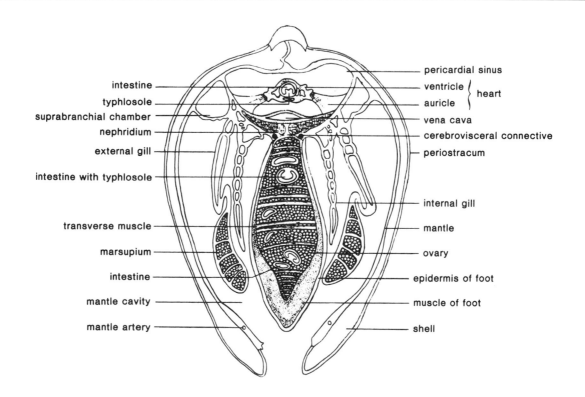

Fig. 11.4 *Freshwater mussel, cross section through heart region.*

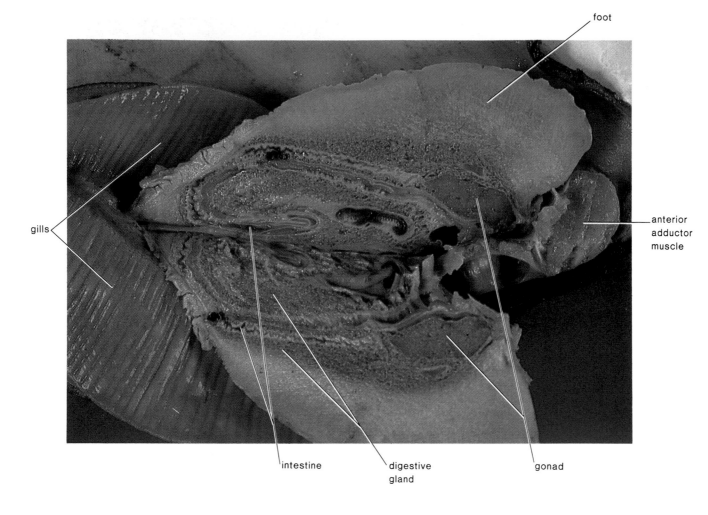

foot

gills

anterior
adductor
muscle

intestine   digestive   gonad
             gland

Fig. 11.5 *Freshwater mussel. Foot dissected to show internal structures. (Photograph by Ken Taylor.)*

If the gills of your clam appear fat or swollen, they may contain eggs or larvae, since the gills of female freshwater clams also serve as a brood pouch or **marsupium** (see figure 11.4). Reproduction and the life history of the freshwater clam will be considered later in this exercise.

Most parts of the digestive system, including the **esophagus, stomach, digestive gland,** and **intestine,** are located within the foot and visceral mass (figure 11.5). To study the various structures enclosed within the **visceral mass,** you will need to cut along the ventral surface of the foot and bisect it into right and left halves. After you have bisected the foot, you should be able to locate the digestive structures mentioned above and the yellowish **gonad** tissue that surrounds a portion of the intestine.

Find the mouth, which is behind the anterior adductor muscle. Food particles, collected on the gills and transported to the labial palps, pass through the mouth and esophagus and into the stomach, where the process of digestion begins. The stomach floor is folded into numerous ciliary grooves that aid in sorting food particles from sediment and other nondigestible particles. Such sorting begins when particulate matter is trapped on the

gills, and it continues enroute to the stomach. Particles rejected in the stomach are passed into the intestine for elimination.

Digestive enzymes are released into the stomach by the **crystalline style,** a gelatinous rod that extends into the stomach and releases enzymes as it is rotated by action of certain stomach muscles. Other enzymes are secreted by the digestive gland. Digestion is both extracellular and intracellular. Small food particles are engulfed by phagocytic cells in the digestive gland, and digestion is completed intracellularly. Undigested material passes through the intestine to the **rectum,** exits through the **anus,** and is flushed from the mantle cavity via the excurrent siphon.

■ Review the location and function of each part of the digestive system and observe how water currents and the gills play a central role in both digestion and respiration. To verify your understanding of the processes of feeding and digestion, trace the path of a group of food and nonfood particles from its entrance through the incurrent siphon to the ejection of the undigested material from the excurrent siphon. Be sure that you know the role of each structure along this path.

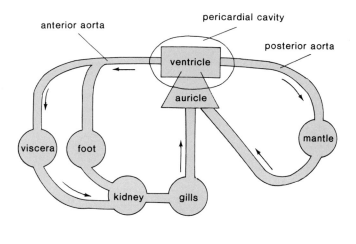

Fig. 11.6 *Pattern of circulation in a mussel.*

Fig. 11.7 *Glochidia larvae of freshwater mussel. (Courtesy of Carolina Biological Supply Company, Burlington, N.C.)*

Muscles

You previously located two large and important muscles. Three other muscles facilitate movements of the foot. The **anterior protractor** aids in extending the foot, and the **anterior foot retractor** draws the foot into the shell. These two muscles are located near the mouth and the anterior adductor muscle. Often they are partially fused with the anterior adductor and may be difficult to distinguish from it.

The **posterior foot retractor** is found near the posterior end of the shell, slightly dorsal and anterior to this larger muscle. The posterior foot retractor draws the foot toward the posterior of the shell.

Circulation

The heart is located in the **pericardial cavity** found near the ventral surface (see figure 11.3). The pericardial sinus is enclosed by a thin membrane, the **pericardium.** The pericardial cavity represents the reduced coelom of the mussel. Locate within the pericardial cavity the muscular **ventricle** surrounding the intestine. Attached to it are two thin-walled **auricles.** Two major blood vessels carry blood away from the heart; the **anterior aorta** supplies the foot and most of the viscera, and the **posterior aorta** supplies the rectum and the mantle.

The circulatory system of the mussel is an **open system.** This means that the blood is not confined to a definite system of closed vessels, but that in the body tissues blood passes into large spaces or sinuses (see figure 11.6). From the visceral organs, blood passes to the nephridia for the removal of metabolic wastes and then to the gills for gas exchange before returning to the heart via the veins. The mantle also serves as a respiratory organ, and oxygen-rich blood from the mantle returns directly to the heart. What color is the blood of the mussel? Does it have hemoglobin?

Excretion, Osmoregulation, and Reproduction

Ventral to the heart and embedded in the mantle tissue is a mass of brownish or greenish tissue, the **nephridia.** The nephridia are the excretory organs, or "kidney," of the mussel. They remove nitrogenous and other wastes from the blood. In freshwater mussels, the nephridia also excrete large amounts of water to maintain proper water balance in the body tissues.

The sexes are separate in freshwater mussels, but the male and female gonads are generally quite similar in appearance. The gonads consist of a yellowish mass of tissue and are located within the foot, surrounding a portion of the intestine (see figure 11.5).

Mature females of some species can be identified easily by the eggs and young larvae contained in the **marsupium,** a specialized portion of the gills that serves as a brood pouch (see figure 11.4). Most clams and mussels shed their gametes into the water where fertilization and embryonic development take place and result in the production of a free-swimming larva. Freshwater mussels (Family Unionidae), however, have a peculiar and highly specialized mode of reproduction.

The ripe eggs are released, and they pass into the suprabranchial chamber, where they are fertilized by sperm brought in by water currents. The fertilized eggs attach to the gills, which enlarge to form the **marsupia** (see figure 11.4). The eggs develop into tiny bivalve larvae, the **glochidia** (figure 11.7), which are released from the mussel to spend a portion of their lives as external parasites on certain species of fish. Later, the parasitic larvae detach from the host fish and move to the bottom, where they become free-living and develop into mature mussels (see demonstration).

Fig. 11.8 Helix, *the garden snail, preserved specimen. (Courtesy of Carolina Biological Supply Company, Burlington, N.C.)*

■ If you have a mature female with a marsupium, cut away a portion of the gill and remove a few of the eggs or larvae. Examine the contents under a stereoscopic microscope. Can you identify the stages of development represented among the eggs and/or larvae?

Nervous Coordination

Although the mussels lack a differentiated head, the basic plan of their nervous system is similar to that of other molluscs and consists of three pairs of ganglia and their connecting nerves. A pair of **cerebropleural ganglia** (right and left) is located just posterior to the anterior adductor muscle; the two **visceral ganglia** lie just ventral to the posterior adductor muscle; and the **pedal ganglia** are embedded in the visceral tissue within the muscular foot.

Demonstrations
1. Living pelecypods in aquaria to show action of siphons.
2. Microscopic section of mussel shell.
3. Glochidium larva.
4. Living or preserved specimens of other common pelecypods.

Helix, the Garden Snail (Demonstration)

Class Gastropoda

Helix is a large land snail often called the garden snail or the edible snail. It is the snail found in gourmet food shops and served in restaurants. Study a specimen that has been relaxed before preservation (figure 11.8) and identify its fleshy anterior **head** with two pairs of **tentacles,** one pair of **eyes** at the tip of the second pair of tentacles, and a ventral **mouth.** Posterior to the head is the large, muscular **foot.** Above the foot is the **visceral mass** covered by the calcareous, coiled shell. When the animal is threatened or disturbed, the head and foot can be retracted into the dorsal shell. Some snails (not *Helix*) have a chitinous plate, the operculum, that covers the opening when the head and foot are withdrawn into the shell.

The anatomy of *Helix* and most other gastropods is difficult to work out because the adult body is coiled and has lost much of the bilateral symmetry of its internal organs. Study a demonstration of a dissected snail to observe the principal internal organs (figure 11.9).

The digestive system consists of the ventral **mouth;** a muscular **pharynx** with a dorsal chitinous **jaw** and a ventral **radula;** a tubular **esophagus;** a large, thin-walled **crop** for storage; and a rounded **stomach.** The **radula** is a chitinous structure used to shred the plant material on which *Helix* feeds. The radula of the *Helix,* like that of many other gastropods, consists of numerous chitinous teeth attached to a muscular organ that can move the radula back and forth against food materials (figure 11.10). The shredded particles are then ingested.

Food passes through the mouth and on to the stomach where it is digested. A **digestive gland** located in the apex of the shell releases enzymes into the stomach. Connecting to the stomach is a long, coiled intestine terminating in the anus, which empties through the fleshy mantle on the right side of the shell.

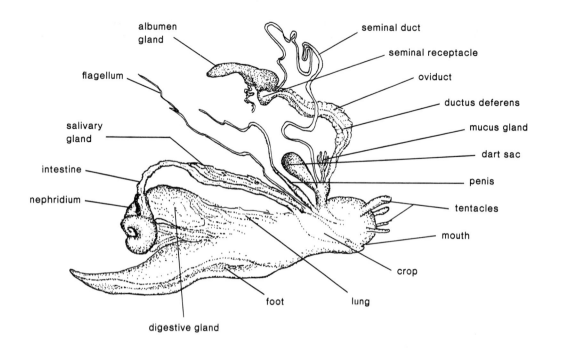

Fig. 11.9 Helix, *the garden snail, internal organs.*

Respiration in *Helix* is accomplished by a highly vascularized portion of the mantle that serves as a **lung.** The circulatory system consists of a **heart** with one **auricle** and one **ventricle.** Blood is pumped from the ventricle through arteries of the principal organs and drains into **blood sinuses** (spaces) from which it flows to the vascularized lung portion of the mantle. A series of pulmonary vessels collect oxygenated blood from the lung and carry it to the heart.

Helix has no real brain, but the nervous system consists of four pairs of **ganglia** connected by nerves with each other and with major organs of the body. The principal sense organs are the two **eyes,** an **olfactory organ,** and two **statocysts,** which serve as balancing organs.

Helix is monoecious. A single, unpaired **gonad,** located near the apex of the shell, produces both eggs and sperm. Mating between two snails results in the exchange of sperm and cross fertilization. Later, each of the snails deposits eggs in a gelatinous mass. Development is direct into miniature snails without larval stages.

Demonstrations
1. Microscope slide of radula.
2. Examples of dextral and sinestral shells.
3. Examples of gastropods with reduced, internal, or no shell (slugs, nudibranchs, etc.).
4. Living or preserved specimens of other gastropods.

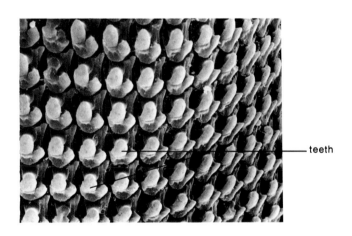

Fig. 11.10 *Radula of snail. Magnification 660×. (Photograph by Louis de Vos.)*

A Squid: *Loligo*

Class Cephalopoda

Squid are large, swift-moving, and highly specialized molluscs (figure 11.11). They exhibit a well-developed nervous system, complex sense organs, and sophisticated patterns of behavior. They are good representatives of the Class Cephalopoda, the most highly evolved class of molluscs. The name cephalopoda means "head-foot" and is quite appropriate because the foot, one of the basic anatomical elements of molluscs, is highly modified in this

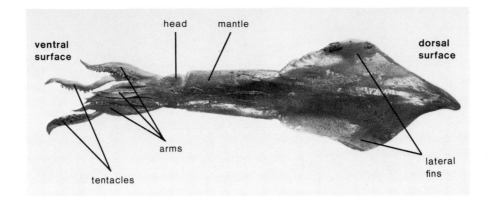

Fig. 11.11 *Squid,* Loligo, *anterior view. (Courtesy of Carolina Biological Supply Company, Burlington, N.C.)*

class and actually serves as a head. The "head-foot" contains several large nerve ganglia, which coordinate nerve impulses and several sense organs that provide important sensory information.

Squid are often abundant in the surface waters of the oceans, where they may form large schools. Dozens or hundreds of squid may be captured together in trawl nets cast overboard in Atlantic and Pacific coastal waters. Other species of squid inhabit the ocean depths.

Several species of the genus *Loligo* are found in the coastal waters of the Atlantic and Pacific Oceans and in the Gulf of Mexico. *Loligo pealeii* is a common small Atlantic squid; *Loligo opalescens* is a small squid abundant in Pacific coastal waters. *Architeuthis,* the giant squid found in deeper waters of the Atlantic Ocean, is the largest known invertebrate animal, with some individuals reaching a length of 15–20 meters.

External Anatomy

■ Examine a preserved squid to identify its major external features. Living squid in an aquarium are interesting to watch and to observe for swimming, feeding, and other types of behavior, but are difficult to obtain and keep in inland locations.

On the preserved specimen observe the **head, tentacles,** and two large **eyes.** The eyes are equipped with a pupil, iris, cornea, lens, and retina and are capable of forming clear images; they are remarkably similar to vertebrate eyes in many respects. This is an interesting and important example of **convergent evolution,** in which two organisms with quite different evolutionary histories (squid and humans) have evolved quite similar structures.

Note that the body of the squid is slender and tapered, which facilitates its swift movement through the water. The orientation of the squid is rather unusual, since morphologically the head and tentacles represent the **ventral surface** of the body and the pointed end opposite the head represents the **dorsal surface** (see figure 11.11). The

funnel or siphon is located on the **posterior surface.** While swimming, a squid actually travels with its dorsal surface forward and its anterior surface up!

Squid swim by a type of jet propulsion in which water is squirted out of the funnel. Since the funnel is under muscular control, the direction of water ejection and thus the direction of movement can be swiftly changed. Movements of the arms and the two lateral fins also aid in steering the animal.

Locate the four pairs of **arms** on the head of your specimen and two pairs of elongated **tentacles.** The tentacles are retractile and play an important role in feeding. Study one arm and observe that it bears two rows of **suckers** (figure 11.12). Are the suckers all the same size? Which suckers are largest? Which are the smallest?

Each sucker is made up of a rounded cup attached to the arm by a stalk or **pedicle.** Remove a sucker from one of the arms and observe it under a stereoscopic microscope. Identify the **chitinous ring** with teeth, which supports the edge of the cup, and the small **piston** at the base of the cup. Note that the suckers on the two tentacles are found only near the distal ends.

A special modification of the suckers on one of the arms of mature males results in a reduction in the size of the suckers and an increase in the length of the pedicle supporting the terminal suckers (figure 11.12). This modified arm is called a **hectocotyl** and serves in the transfer of sperm bundles (spermatophores) to the female in mating. Such a modification of an anatomical structure by one sex (males in this case) and not the other is an example of **sexual dimorphism.**

Locate the muscular membrane at the base of the arms and tentacles. In the center of the membrane is the **mouth** opening (see figure 11.12). Inside the mouth find the two chitinous **beaks,** which are used to seize and tear the prey in feeding. Prey organisms are captured by swift movements of tentacles, which seize the prey, hold it fast with the suckers, and draw it toward the mouth with the aid of the other arms. Poison glands provide secretions which aid in subduing active prey.

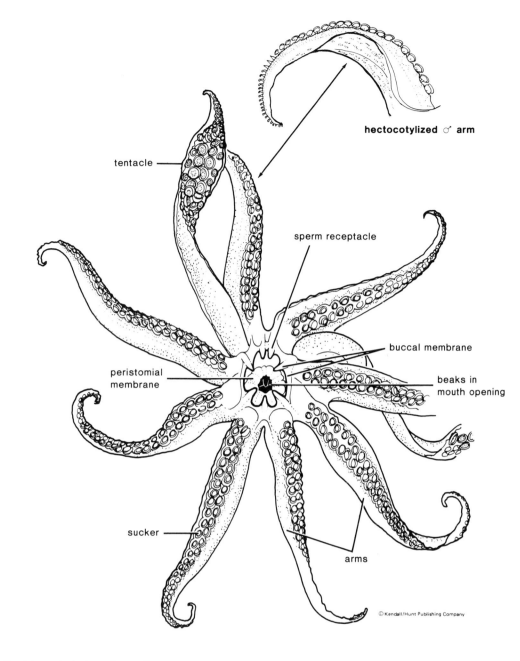

hectocotylized ♂ arm

tentacle

sperm receptacle

buccal membrane

peristomial membrane

beaks in mouth opening

sucker

arms

© Kendall/Hunt Publishing Company

Fig. 11.12 *Squid, ventral view of mouth area.*

The fleshy mantle surrounds the remainder of the body and encloses the internal organs. The ventral edge of the mantle adjacent to the eyes is called the **collar.** Locate the **funnel** extending from the posterior side of the collar. On the anterior side of the collar, you should find a projection of the mantle that represents the ventral tip of the pen. In the squid, the skeleton is reduced to a translucent, internal, chitinous structure called the **pen** plus several cartilages, which protect the **cephalic ganglia** or "brain" within the head-foot, and other cartilages embedded in the collar region of the mantle and funnel.

Locate also the two large **lateral fins** extending from the mantle. As noted above, the fins are movable and serve as steering aids during swimming. On the surface of the mantle are numerous **chromatophores,** specialized pigment-containing cells that allow the squid to change the color of its integument. A chromatophore consists of a small sac of pigment surrounded by an elastic membrane attached to many small muscles. Contraction of these tiny muscles causes the pigment to spread out and add to the general coloration of the squid. A chromatophore whose pigment is not spread out contributes little to the apparent color of the squid.

Since squid have several types of chromatophores, each with a different colored pigment, many shades of color can be achieved. This aids greatly in the camouflage and protective coloration of the squid. Live squid show almost

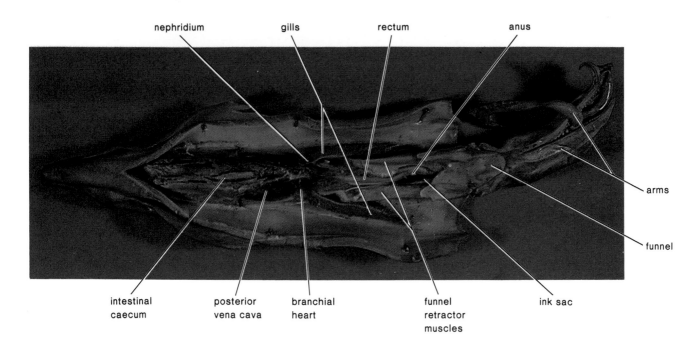

nephridium   gills   rectum   anus

arms

funnel

intestinal   posterior   branchial   funnel   ink sac
caecum   vena cava   heart   retractor
    muscles

Fig. 11.13 *Squid. Dissection of male, posterior view. (Photograph by Ken Taylor.)*

constant changes of color because of the continuous activity of the chromatophores, which are under nervous (and muscular) control.

Internal Anatomy (Demonstration)

Study a demonstration of a dissected specimen to observe the principal organs (figure 11.13). Posterior to the head is the **funnel** connected with two large **funnel retractor muscles**. Dorsal to the funnel, locate the **rectum** and **ink sac**. A median **kidney** surrounds the **intestine**. Two **branchial hearts** pump blood to the two **gills**. Anterior to the paired branchial hearts is the **intestinal caecum**. A chitinous **pen** is attached to the anterior wall of the mantle cavity and provides support for the body.

Octopus (Demonstration)

Class Cephalopoda

The body of an octopus is globular and saclike with well-developed eyes and a complex brain (figure 11.14). Octopi usually crawl about the sea bottom using their eight flexible arms equipped with numerous suction cups for locomotion and to grasp prey. They can also swim by jets of water forced through the funnel from the mantle cavity when startled.

Octopi are carnivores and feed on molluscs by drilling holes in the shells with their radula and injecting a poison to kill their prey. They also have a powerful chitinous beak by means of which they can shred tissue or attack enemies.

Demonstrations

1. Dried pen of squid to illustrate internal skeleton.
2. Preserved or plastic mount of octopus.
3. Shell of chambered nautilus.

Key Terms

Adaptive radiation evolutionary diversification of a group of animals with common ancestry to fill numerous ecological niches.

Chromatophores specialized pigment-containing cells found in the integument of cephalopod molluscs and many crustaceans (Phylum Arthropoda).

Coelom central body cavity lined by mesodermal tissues; develops from spaces arising within the mesoderm during embryonic development. Characteristic of annelids, molluscs, and all higher metazoans, although modern molluscs have a secondarily reduced coelom.

arms

eye

suckers

mouth

Fig. 11.14 *Octopus. (Courtesy of Carolina Biological Supply Company, Burlington, N.C.)*

Convergent evolution the evolution of two apparently similar structures by two or more groups of animals with different ancestral lines.

Foot in molluscs, the foot is a large ventral mass of muscular tissue; usually functions in locomotion; sometimes modified for other functions.

Glochidium larva specialized larval form in the life cycle of certain freshwater pelecypod molluscs; enclosed in a transparent bivalve shell, these larvae spend part of their life cycle as parasites on the gills of freshwater fish. Later they detach and develop into free-living adult mussels. Plural: Glochidia.

Hemocyanin a blue copper-containing respiratory pigment in the blood of molluscs and certain other invertebrate animals. Carries oxygen to the body cells.

Mantle a thin outer layer of tissue largely enveloping the bodies of most molluscs. Usually secretes a shell.

Marsupium a modified portion of the gills in females of certain freshwater clams and mussels; serves as a brood pouch for eggs and young larvae.

Sexual dimorphism condition in which male and female members exhibit a different form.

Notes and Sketches

12

Annelida

Objectives

After completing the laboratory work in this chapter, you should be able to perform the following tasks:

1. Briefly outline the characteristics of the Phylum Annelida and cite five significant advances over the Phylum Platyhelminthes.

2. List and briefly characterize the three classes of Annelids. Cite an example of each.

3. Identify the principal external features of the polychaete worm *Nereis* and explain their functions.

4. Describe the main features of the external anatomy of the earthworm and tell how the earthworm exhibits morphological adaptations for life in the soil.

5. Identify the main structures seen in a microscopic cross section of an earthworm and explain the function of each part.

6. Identify the parts of the digestive system in a dissected earthworm and give the function of each part.

7. Describe the basic pattern of circulation and identify the chief parts of the circulatory system of an earthworm.

8. Identify the main reproductive organs and discuss reproduction in earthworms.

9. Describe the external anatomy of a leech and tell how it differs from that of an earthworm.

Introduction

The Phylum Annelida includes more than 9,000 species of segmented worms, animals whose cylindrical, elongate bodies are comprised of a series of longitudinal segments (also called somites or metameres). In addition to their conspicuous segmentation, the annelids also exhibit several other important structural advances over the flatworms and roundworms that have been studied previously. Some important advances include: a spacious coelom, a closed circulatory system, an efficient excretory system, a highly developed muscular system, a well-developed central nervous system, and an increased concentration of nerve centers (ganglia) and sensory organs at the anterior end of the animal (cephalization).

Annelids are the first animals we have studied that exhibit a well-developed coelom. Although molluscs are also coelomate animals, the coelom of molluscs is reduced to the cavities in which the heart (pericardial cavity), nephridia, and gonads are located. The annelids, in contrast, have a large coelomic cavity in which most of the internal organs are suspended. Annelid worms illustrate well the basic body organization of a coelomate animal and the relationship of the coelomic cavity, its peritoneal lining of mesodermal origin, and the internal organs.

Classification

The phylum is divided into three classes, chiefly on the basis of **segmentation,** distribution of setae, and the presence or absence of a clitellum. Setae are thin, chitinous bristles or rods that are secreted by certain cells in the body wall. Setae are used in locomotion and, in many annelids, also aid in feeding. The number, morphology, and distribution of setae is often important in the classification of annelids (figure 12.1).

The clitellum is a glandular swelling formed seasonally by certain segments. It secretes a cocoon into which eggs are deposited for fertilization and later incubation.

Class Polychaeta (Polychaete Worms)

Annelid worms with many segments, lateral appendages (parapodia) with many setae, a well-developed head region, no clitellum, lacking permanent gonads, usually dioecious, and typically having a trochophore larva. Most species are marine. This class also includes the archiannelids, a group of small marine worms formerly considered by many biologists to be a separate class. Examples: *Nereis* (formerly *Neanthes,* the clamworm); *Arenicola* (lug worm); *Chaetopterus; Hydroides* (feather duster or fan worm); *Diopatra* (plume worm); *Polygordius* (an archiannelid).

Class Oligochaeta (Bristleworms)

This class includes the earthworms and many species of related freshwater worms that have many segments but few setae per segment, form a clitellum, but lack parapodia and a differentiated head region. They are generally

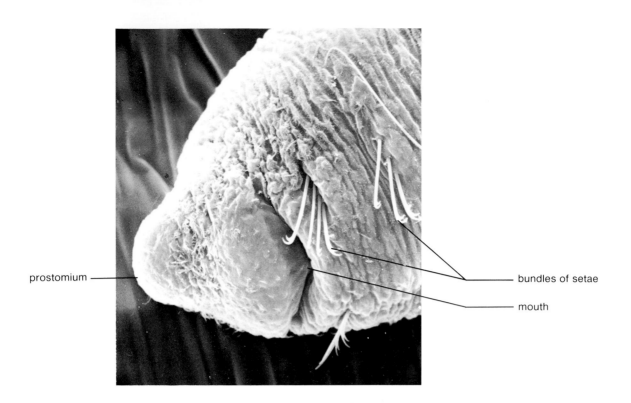

prostomium

bundles of setae

mouth

Fig. 12.1 *Mouth region of an aquatic oligochaete showing characteristic arrangement and morphology of setae. Magnification 450X. (Scanning electron micrograph by Thomas Bouillon.)*

monoecious, have permanent gonads, and development is direct. Mostly found in the soil and in freshwater habitats. Examples: *Lumbricus* and *Allolobophora* (earthworms); *Eisenia* (dung worm); *Tubifex* (sewage worm); *Aeolosoma; Chaetogaster*. The last three are aquatic genera and are common freshwater forms.

Class Hirudinea (Leeches)

This group includes the leeches, a relatively small class (about 500 species) of specialized annelids found mainly in freshwater habitats. A few species are also terrestrial. Leeches have dorsoventrally flattened bodies, anterior and posterior suckers, thick muscular bodies, no head or tentacles at their anterior ends, and lack parapodia and setae. The number of segments is limited (34) and a clitellum is formed during the reproductive season.

Three representatives of this phylum have been selected for special study to illustrate the principal features of the Annelida: the clamworm, an earthworm, and a leech.

Materials List

Living specimens
    Earthworm
Preserved specimens
    Nereis
    Leech
    Representative polychaetes (Demonstration)
    Polychaete tubes (Demonstration)
    Earthworm cocoons (Demonstration)
    Representative oligochaetes (Demonstration)
    Representative leeches (Demonstration)
    Leech, dissected to show internal anatomy
    (Demonstration)
Prepared microscope slides
    Earthworm, cross section
    Trochophore larva (Demonstration)
Miscellaneous supplies
    Dissecting pans
    Pins
Audiovisual materials
    Chart of earthworm anatomy

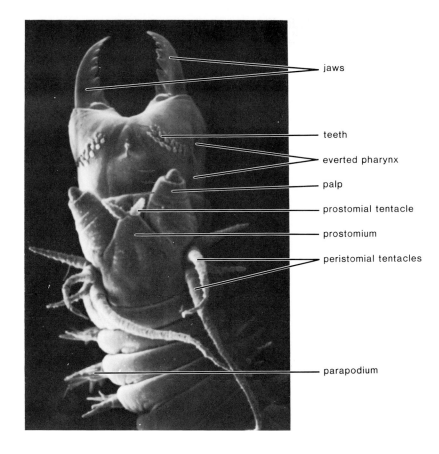

jaws

teeth

everted pharynx

palp

prostomial tentacle

prostomium

peristomial tentacles

parapodium

Fig. 12.2 Nereis, *head region, dorsal view. (Scanning electron micrograph by Betsy Brown.)*

A Marine Annelid: *Nereis virens*

Class Polychaeta

The sandworm or clamworm, *Nereis virens* (figures 12.2 and 12.3), is a large marine annelid commonly found in many areas along the Atlantic coast of North America. *Nereis* usually lives in burrows in the sand or in mud bottoms of the shallow coastal waters during the day and emerges at night to search for food.

■ Select a preserved specimen of *Nereis* and observe the numerous body segments. Internally, the segments are separated by thin sheets of tissue called **septa.** The internal structure of *Nereis* is generally similar to that of the earthworm, which you will study later.

All the segments of *Nereis* except the first and the last bear a pair of lateral **parapodia** (see figure 12.4). Observe the demonstration slide of a cross section of *Nereis* to study the structure of these appendages. Note that the parapodia are **biramous** (two-branched), with a dorsal **notopodium** and a ventral **neuropodium.** Each main branch bears bundles of bristles, or **setae,** and also has several smaller side branches. The parapodia play an important role both in locomotion (swimming and creeping along the sea bottom) and in respiration. What aspects of their structure appear to be advantageous for these functions?

Examine the well-developed head region made up of the **peristomium,** the first complete body segment that encircles the mouth, and the **prostomium,** a triangular mass of tissue located on top of the peristomium (figure 12.2). Several sensory structures are found on the prostomium, including two fleshy **palps** attached to the sides of the prostomium, two pairs of pigmented **eyes** on the dorsal surface, and two **anterior tentacles.** Four pairs of **lateral tentacles** (or cirri) are also found attached to the peristomium.

During feeding, the pharynx may be everted through the mouth, exposing two powerful **jaws** and many small **teeth** on the wall of the pharynx (figures 12.2 and 12.3).

Fig. 12.3   Nereis, *head region, ventral view. (Scanning electron micrograph by Betsy Brown.)*

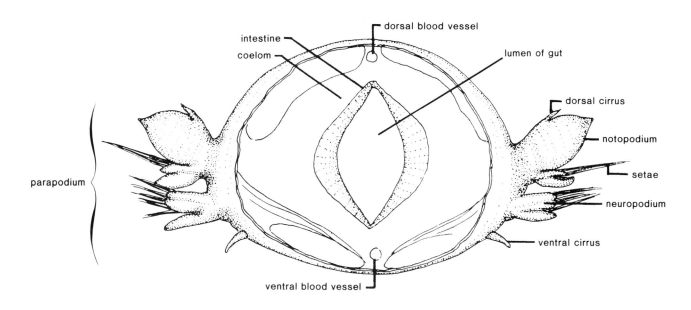

Fig. 12.4   Nereis, *cross section.*

The jaws aid in seizing prey, and the pharyngeal teeth help hold the prey as the pharynx is withdrawn through the mouth.

The sexes are separate in *Nereis,* as in most polychaete worms, although there are no permanent gonads. The gametes (eggs and sperm) are formed in the body wall at certain times, released into the coelom, and then released into the sea. There the eggs may be fertilized and develop into a **trochophore larva** (figure 12.5). Other larval stages follow before metamorphosis into the adult worm.

Many species of polychaete worms show marked structural, physiological, and behavioral changes during the breeding season. Some species aggregate in large swarms and rise to the surface before releasing their gametes into the sea. This special behavior represents an important adaptation that serves to increase the likelihood of fertilization and the continuation of the species.

The trochophore larva formed by many marine polychaetes bears a close resemblance to the larvae of many marine molluscs. This similarity of trochophore larvae is one of the important pieces of evidence suggesting a close evolutionary relationship between these two phyla. Similarly, the resemblance of the trochophore larva of annelids and molluscs to the marine larvae of several other invertebrate phyla, plus other similarities in their patterns of embryonic and larval development, suggests other important evolutionary relationships with such groups as the flatworms and the arthropods.

Other Polychaetes

Polychaete worms are often abundant in the shallow coastal waters of the sea. This class includes many varied forms, some of which form permanent tubes that they secrete or build from various materials (often called sedentary polychaetes). Others do not form permanent tubes (called errant polychaetes). Observe the demonstrations of various living and preserved polychaetes available in the laboratory. Three common and well-known species are illustrated in figures 12.6, 12.7, and 12.8.

Hydroides (figure 12.6) secretes a permanent calcareous tube on the surface of shells and rocks. The tube is open at the anterior end and tapers to the closed posterior end. While feeding, the worm extends a feathery crown of anterior appendages (extensions of the prostomium) from the open anterior end of its calcareous tube. Because of these feathery appendages, which are often brightly colored, *Hydroides* and other members of the Family Serpulidae are called feather duster or fan worms.

Diopatra, the plume worm (figure 12.7), is a polychaete common in intertidal areas on the Atlantic coast of the United States. The body of the plume worm is iridescent and bears several pairs of bright red (blood-filled)

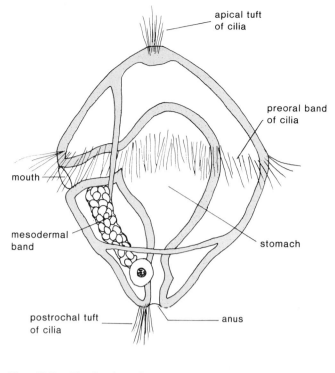

Fig. 12.5 *Trochophore larva.*

gills. *Diopatra* forms a vertical, parchmentlike tube which may reach three feet in length. Most of the tube is buried in the sediment, but 2–3 inches usually extend above the sediment. The tube is camouflaged by bits of shell, sea grass, and other bits of debris.

Arenicola, the lug worm (figure 12.8), is a burrowing polychaete which lives in a J-shaped tube lined with mucus. In its habits *Arenicola* resembles a seagoing earthworm. The worm burrows in shallow marine sediments by extending its proboscis into the sand or mud and retracting it filled with sand. The sand contains detritus and plankton on which the worm feeds as the sand passes through the digestive tract. *Arenicola* shows several external structural adaptations related to its peculiar burrowing habits, including reduced parapodia, small gills, a large bulbous proboscis, and the lack of tentacles and other head appendages.

Demonstrations

1. Representative species of Polychaeta, such as *Amphitrite, Aphrodite, Arenicola, Chaetopterus, Hydroides, Sabella, Diopatra,* and *Polygordius.*
2. Examples of different kinds of tubes formed by polychaetes.
3. Microscope slide of trochophore larva.

Fig. 12.6 Hydroides, *a fan worm.*

operculum

calcareous tube

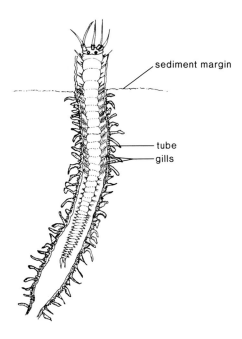

sediment margin

tube
gills

Fig. 12.7 Diopatra, *the plume worm.*

Fig. 12.8 Arenicola, *the lug worm.*

The Earthworm

Class Oligochaeta

Earthworms are terrestrial annelids that typically live in tunnels they burrow in the soil. Internally, they demonstrate many of the typical features of the Phylum Annelida, and they are the most commonly studied representatives of the phylum. The external features of earthworms, however, exhibit several modifications apparently related to their burrowing habits and underground life. Most apparent of these external adaptations are the absence of a distinct head and the lack of parapodia.

External Anatomy

■ Obtain an anesthetized or freshly killed specimen of a large species of earthworm and examine some of the major external features. Preserved specimens may also be studied, but fresh specimens are much superior.

Observe the long, cylindrical body, which tapers at each end. The anterior end can be identified most readily by reference to the **clitellum,** a conspicuous swollen region including several segments located near the anterior end. The clitellum secretes a **cocoon** in which the fertilized eggs are incubated (figure 12.9). Note the absence of a distinct head and the lack of external sense organs at the anterior end.

The **mouth** is located in the first segment, and a small rounded projection, the **prostomium,** overhangs the mouth. The **anus** is located in the last segment.

Observe the darker and more rounded dorsal surface. Locate the **setae,** small chitinous bristles in each segment except the first and the last. To find the setae, rub your finger lightly back and forth along the sides of the worm. Observe how the setae protrude slightly from the body surface and catch on your finger. The setae are located in four distinct rows. Two pairs of setae are located in rows on the ventral surface, and two pairs are located on the sides (see figure 12.10). Toward which end of the worm do the setae point? What is the advantage of this orientation in locomotion?

Note the iridescent **cuticle,** a protective covering which restricts water loss, that covers the surface of the body. The iridescence results from the many small striations in the surface of the cuticle. These striations cause the cuticle to act as a diffraction grating to separate incident light into separate wavelengths and produce the iridescence.

Also located on the surface of the body are several openings from the excretory and reproductive systems. A pair of small **excretory pores,** the nephridiopores, is found on the ventral surface of each segment except the first three or four and the last. Openings from the reproductive

Fig. 12.9 *Earthworm cocoons. (Courtesy of Carolina Biological Supply Company, Burlington, N.C.)*

system include: a pair of **male pores** on the ventral surface of segment XV surrounded by swollen "lips," a small pair of **female pores** on the ventral surface of segment XIV, and openings to two pairs of **seminal receptacles** located laterally in the grooves between segments IX–X and X–XI. Notice that the segments are usually designated by Roman numerals. The use of a stereoscopic (dissecting) microscope will help in locating these tiny external openings to the excretory and reproductive systems of the earthworm.

Note: The location of these pores and the internal reproductive organs differs in various species of earthworms. The traditional earthworm for dissection is *Lumbricus terrestris,* upon which this and most other textbook descriptions are based. Many of the earthworms supplied for dissection by American biological supply houses, however, are different species and therefore differ in some minor respects from textbook descriptions, particularly in respect to the location of the reproductive structures in particular segments.

Internal Anatomy

■ Place your earthworm in a wax-bottom dissecting pan and carefully pin the specimen down, dorsal side up, with one pin through the prostomium and another near the posterior end. Study figure 12.14 and note the thin body wall and the closeness of the underlying parts before you begin your dissection. Puncture the body wall slightly to one side of the dorsal midline with the tips of your scissors and make a longitudinal incision from behind the clitellum, continuing it anteriorly to the mouth. Cut to the side of the large dorsal blood vessel and avoid severing it. Take special care, also, to avoid damage to the brain, which is located near the mouth.

Fig. 12.10 *Earthworm portion of body wall showing lateral rows of setae. Magnification 50×. (Scanning electron micrograph by Louis de Vos.)*

Fig. 12.11 *Earthworm, dissected, dorsal view. (Photograph by Ken Taylor.)*

Figure 12.11 shows a dissected earthworm and its internal organs. Observe the large body cavity, the **coelom,** that is divided into many distinct compartments by the **septa.** The septa are membranous, transverse partitions that divide the body of the earthworm internally into many segments.

Carefully separate the body wall from the internal structures and pin down the body wall on each side. Place the first pins on opposite sides of segment XV, easily located because of the presence of the large male pores. Insert the pins through the tissue and into the wax bottom of the pan at a 45° angle to provide working space and a

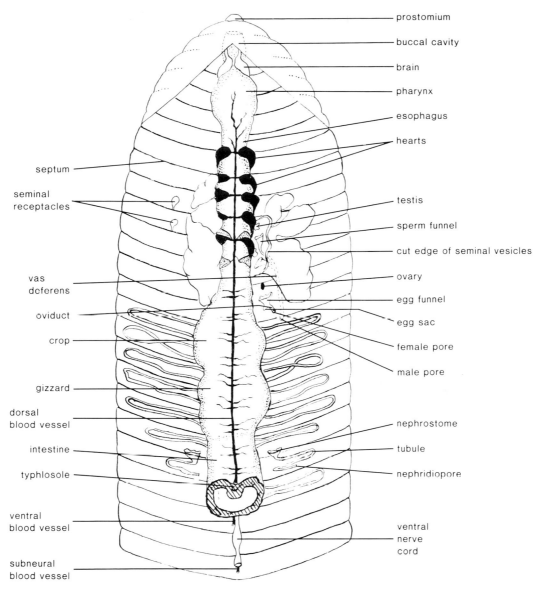

prostomium
buccal cavity
brain
pharynx
esophagus
hearts
testis
sperm funnel
cut edge of seminal vesicles
ovary
egg funnel
egg sac
female pore
male pore
nephrostome
tubule
nephridiopore
ventral nerve cord

septum
seminal receptacles
vas deferens
oviduct
crop
gizzard
dorsal blood vessel
intestine
typhlosole
ventral blood vessel
subneural blood vessel

Fig. 12.12 *Earthworm, principal internal organs, dorsal view.*

clear view of the internal organs. Place additional pins through each fifth segment, for example, segments V, X, XV, XX, XXV, XXX, etc., to serve as convenient landmarks to identify the location of various structures. Cover the specimen with about one centimeter of water to keep the internal organs moist and flexible.

Digestive System

The digestive tract is a straight tube extending from the mouth to the anus. Examine your specimen and locate the **mouth** at the anterior end. The mouth opens into the small **buccal cavity** (see figure 12.12). Immediately behind the buccal cavity is the muscular **pharynx.** The pharynx is at-

tached to the body wall by numerous threadlike **dilator muscles** that can produce a sucking action to draw food materials through the mouth into the buccal cavity and pharynx. The tubular, thin-walled **esophagus** extends from segments VI to XIII and passes the food posteriorly to the **crop.**

Associated with the lateral wall of the esophagus of earthworms are two or more pairs of **calciferous glands.** These glands develop as evaginations of the esophageal wall; they function as excretory rather than digestive organs. The calciferous glands serve to regulate the levels of calcium and carbonate ions (Ca^{++} and $CO_3^{=}$) and the pH of the blood.

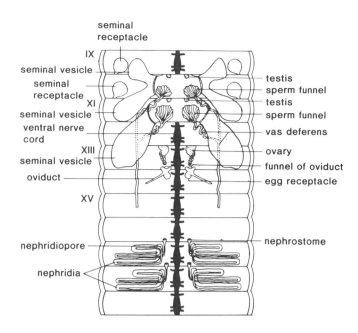

Fig. 12.13 *Earthworm, reproductive and excretory systems.*

Food is stored in the thin-walled, extensible crop before passing into the muscular **gizzard,** where it is ground into smaller pieces and passed to the **intestine** to be digested and absorbed. Undigested materials are passed to the **anus** for egestion.

Much of the intestine, as well as the dorsal blood vessel, is covered by a yellowish layer of **chlorogogue tissue.** The chlorogogue tissue does not appear to be directly involved in digestion, but it does store glycogen and lipids and has other functions analogous to the liver in vertebrate animals, including chemical defense of the body.

Circulatory System

The earthworm has a closed circulatory system, which contains red blood. The red color is due to the respiratory pigment, **hemoglobin,** dissolved in the **plasma,** or fluid portion, of the blood. In addition to the plasma, the blood also contains numerous colorless **blood cells,** or corpuscles.

Locate the large **dorsal blood vessel** lying just above the digestive tract, and also a pair of **parietal vessels** extending laterally from the dorsal blood vessel in each segment. In segments VII to XI, inclusive, the parietal vessels are enlarged and much more muscular. These five pairs of enlarged **aortic arches** are often called "hearts." Their contractions aid in propelling the blood through the circulatory system.

The yellowish tissue that covers the surface of the large blood vessels consists of **chlorogogue cells.** Some of the other major blood vessels include the **ventral vessel,** lying beneath the digestive tract; the **subneural vessel,** lying beneath the ventral nerve cord; and two **lateral neural vessels,** one on each side of the ventral nerve cord (see figure 12.13).

In an anesthetized worm, waves of muscular contraction can be readily observed along the dorsal vessel and the five pairs of "hearts." In which direction does the blood flow through these vessels? Valves in the dorsal vessel and in the hearts serve to prevent backflow and to aid in the efficient circulation of the blood.

Excretory Organs

Each segment, except the first three or four and the last, contains a pair of tubular excretory structures, the **nephridia.** Each nephridium is located partly in two adjacent segments. Consult figures 12.12 and 12.13. Locate the funnel-shaped opening into the coelom, the **nephrostome,** which projects through the septum into the next anterior segment. Locate also the **coiled tubular portion,** which empties through a ventral opening in the body wall, the **nephridiopore.**

The **calciferous glands,** formed as evaginations of the wall of the esophagus, also serve as excretory organs. They were described previously during our discussion of the digestive system.

Reproductive System

■ Cut across the digestive tract about a half-inch behind the gizzard and carefully free it with the associated blood vessels from the underlying structures up to the level of segment IV.

The earthworm is monoecious; each individual has both male and female sex organs (see figure 12.13). Despite the monoecious condition, however, cross fertilization between different individuals is necessary for sexual reproduction in earthworms.

Several of the male reproductive organs are larger and more conspicuous than the female organs. Locate the following male reproductive structures and note the segments in which they are located: (1) three pairs of large **seminal vesicles** (storage sacs in which the sperm mature and are stored until copulation); (2) two pairs of small **testes** embedded within the tissues of the seminal vesicles; (3) two pairs of small ciliated **sperm funnels** (one pair on each side), which connect via short tubules to (4) two **vasa deferentia** (sperm ducts), which lead to the male genital pores on segment XV.

The female reproductive structures are small and difficult to locate in most specimens. They include a pair of **ovaries** in segment XII, which discharge mature ova into the coelom. The ova are collected in two ciliated **egg funnels** and passed through the oviducts to the **female genital openings** on segment XIV. You should be able to locate two pairs of **seminal receptacles** in segments IX and X, where sperm is received during copulation. Sperm stored in the seminal receptacles is later used in fertilization of eggs released into the cocoon.

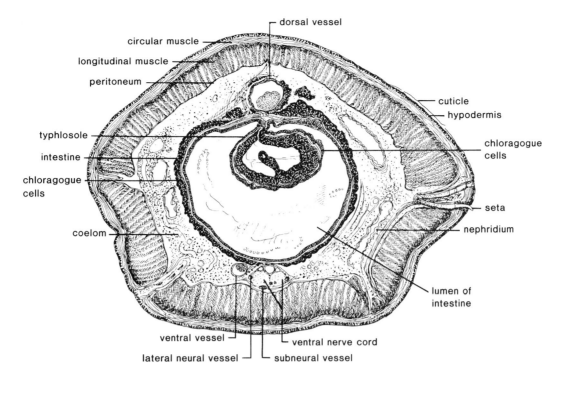

Fig. 12.14 *Earthworm, cross section.*

Nervous System

The nervous system of the earthworm consists of the following principal components: (1) an anterior pair of large **suprapharyngeal ganglia** (the "brain"); (2) two **circumpharyngeal connectives;** (3) a pair of smaller **subpharyngeal ganglia;** and (4) a **ventral nerve cord** with an enlarged ganglion in each segment (see figure 12.13). Locate several nerves extending from the suprapharyngeal and subpharyngeal ganglia to the anterior segments. Three pairs of **lateral nerves**—two from the ganglion itself and one anterior to the ganglion—lead from the ventral nerve cord in each posterior segment.

Cross Sections

■ Obtain a slide with a stained cross section of an earthworm and study its histological structure (figure 12.14). Observe the body wall and its composition. Locate the outer **cuticle** and the underlying **hypodermis.** What is the nature of the cuticle and how is it formed? Beneath the hypodermis is a thin layer of **circular muscles** and a thicker layer of **longitudinal muscles.**

Observe the thin layer of flattened cells, the **peritoneum,** at the inner margin of the longitudinal muscle layer. Covering the outer surface of the intestine (and some of the larger blood vessels as noted earlier) is a specialized type of peritoneum, the **chlorogogue.** The cells of this layer are believed to function like the vertebrate liver in inter-mediary metabolism, in the removal of waste products, and also in the synthesis of hemoglobin, the red oxygen-carrying pigment in the blood. The body cavity of the earthworm is a true **coelom.** What makes this a coelom?

You may be able to observe within the coelom the **nephridia** lateral to the intestine. The nephridia in cross section may appear to be irregularly shaped since they are often cut through the coiled portions of the tubule (figure 12.15).

The wall of the intestine consists of several layers of cells, including an outer **chlorogogue,** thin layers of **longitudinal muscle** and **circular muscle,** and a layer of **endodermal epithelium** lining the cavity, or lumen, of the gut. Dorsally, observe the pronounced fold of tissue, the **typhlosole,** which extends into the lumen. What is the function of the typhlosole?

Observe also the **dorsal** and **ventral blood vessels** and the **ventral nerve cord.** On each side of the nerve cord is a **lateral neural blood vessel,** and beneath the nerve cord is the **subneural blood vessel.**

Demonstrations

1. Charts and models to illustrate earthworm anatomy.
2. Earthworm cocoons.
3. Behavior of living earthworms (locomotion, reactions to stimuli, etc.).
4. Examples of other Oligochaetes, such as *Tubifex, Aeolosoma, Enchytraeus, Stylaria, Chaetogaster.*

Fig. 12.15 *Earthworm, portion of cross section showing opening of nephridiopore through body wall. (Courtesy of Carolina Biological Supply Company, Burlington, N.C.)*

Leeches

Class Hirudinea

Although they are commonly called "bloodsuckers," many of the 500-plus species of leeches (figure 12.16) are actually scavengers or predators on small invertebrates, rather than ectoparasites that feed on the blood of some vertebrate host to which they periodically attach.

The large European leech *Hirudo medicinalis* described in most textbooks was used in the early days of medical practice for bloodletting, since it was once believed that many diseases and bodily disorders were caused by an accumulation of excess blood. For this reason, this species is often called the "medicinal leech."

Most leeches live in freshwater streams, ponds, and lakes. There are also some marine species and a few terrestrial forms. Leeches are relatively large annelids, and the adults of most species reach about 2–5 cm in length. Among their distinctive features is the **lack of appendages and setae,** but the presence of large **anterior and posterior suckers,** a **fixed number of segments,** and male and female **copulatory organs.** The leeches most often available for laboratory study in the United States are large aquatic species of the genera *Haemopis, Malacobdella,* or *Placobdella.*

External Anatomy

■ Examine a preserved leech and observe the elongate, flattened body. The body is flattened dorsoventrally and usually tapers toward the anterior end. Note the absence of setae and lateral appendages. Locate the numerous rings or **annuli** that circle the body.

Leeches are segmented both internally and externally; but, unlike the oligochaetes and polychaetes, leeches have a limited number of body segments (34). Also, the external annuli are more numerous than the internal segments; various internal segments bear from 1–5 annuli each.

Chapter 12

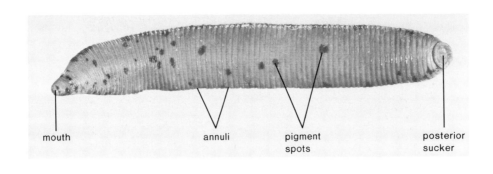

mouth   annuli   pigment spots   posterior sucker

Fig. 12.16 *Leech. (Courtesy of Carolina Biological Supply Company, Burlington, N.C.)*

Identify the large **anterior sucker** and the smaller **posterior sucker.** The suckers serve both for attachment to the host and for locomotion. Leeches move in a looping fashion by attaching to the substrate by the anterior sucker, contracting the body, attaching by the posterior sucker, loosening the anterior sucker, and extending the body forward. Aquatic leeches can also swim by means of undulatory motions of the body resulting from rhythmic contractions of their muscular bodies.

Locate the **mouth** in the center of the anterior sucker. At the anterior margin of the posterior sucker find the **anus** on the middorsal surface.

Internal Anatomy (Demonstration)

■ Study a demonstration of a dissected leech to observe its internal anatomy.

The principal internal features include a complete digestive system with a **mouth** containing teeth, a muscular **pharynx,** a large **crop** that serves to store blood in parasitic leeches, a small **stomach,** a narrow tubular **intestine,** and an enlarged **rectum** that opens dorsally via the **anus** located just anterior to the posterior sucker.

Leeches also have a closed circulatory system, several pairs of segmental nephridia that serve as excretory organs, and permanent gonads. The **reduced coelom** contains one pair of ovaries and several pairs of testes. Although leeches are monoecious (hermaphroditic), reproduction involves copulation and reciprocal fertilization between different individuals. The fertilized eggs are deposited in a cocoon secreted by the clitellum formed by an enlargement of segments 9–11. The eggs develop directly into miniature leeches; there are no larval stages.

Demonstrations

1. Preserved specimens of several different leeches, such as *Haemopis, Malacobdella, Placobdella,* and *Hirudo.*
2. Dissected leech to illustrate internal anatomy.

Key Terms

Biramous having two branches or rami; as the parapodium of a polychaete worm with a dorsal notopodium and a ventral neuropodium.

Clitellum a thickened, glandular portion of certain midbody segments of many oligochaetes and leeches; characteristic of sexually mature individuals. Forms a cocoon in which eggs are deposited and incubated.

Coelom a central body cavity fully lined by peritoneum, an epithelial lining of mesodermal origin. It forms the principal body cavity of annelids and most higher phyla, although in some groups the coelom is secondarily reduced.

Nephridium a tubular excretory organ found in annelids and other invertebrate phyla; may function both in osmoregulation and in the removal of nitrogenous and other body wastes. Plural: nephridia.

Parapodium a lateral appendage typical of polychaete worms; always occurs in pairs on opposite sides of the body; usually with many setae arranged in bundles. Plural: parapodia.

Segmentation having a body consisting of many similar units or subdivisions arranged along the anterior-posterior axis. Each unit is called a segment, somite, or metamere. Also called metamerism.

Septum a thin wall of tissue separating two adjacent segments or masses of tissue. Plural: septa.

Seta a chitinous bristle or rod secreted by certain epidermal cells of oligochaetes and polychaetes. Setae are also found in certain arthropods and other invertebrates. Plural: setae.

Trochophore larva a marine larval form characteristic of many polychaete worms and certain other invertebrate groups; with a pear-shaped body, one or more bands of cilia around the body, tufts of cilia at the anterior and posterior ends, and a complete digestive tract. Similar larvae are formed by molluscs and certain other invertebrate phyla.

13

Arthropoda

Objectives

After completing the laboratory work for this chapter, you should be able to complete the following tasks:

1. List the four subphyla of arthropods. Briefly characterize each subphylum and cite examples.
2. Identify the principal external features of the horseshoe crab, *Limulus;* briefly describe its feeding and locomotion.
3. Demonstrate the chief external features of a spider on a preserved specimen.
4. Compare the basic morphological organization of a spider with that of the horseshoe crab, a crayfish, and a grasshopper.
5. Identify the chief external features of a crayfish. Compare and contrast the morphology of the crayfish with that of the horseshoe crab. Tell which animal is the more primitive and explain why.
6. Discuss the concept of serial homology and give examples from the appendages of a crayfish.
7. Describe the respiratory system of a crayfish and tell how the appendages are involved in its respiration.
8. Describe the circulatory system of a crayfish and compare it with that of an earthworm.
9. Identify the main morphological features of *Daphnia* and explain how this animal is adapted for its planktonic life.
10. Describe the feeding mechanism of *Daphnia* and identify the chief structures involved.
11. Identify the principal external features of the grasshopper *Romalea* and tell how they illustrate the basic organization of the insects.
12. Describe the special adaptations of the appendages of the honeybee and demonstrate them on a specimen.
13. Describe the principal types of insect development and identify the stages in complete metamorphosis.

Introduction

The arthropods represent the largest and most successful of the animal phyla. More than 870,000 species of arthropods have now been described, and members of this phylum dwell in almost every type of habitat. In addition to the many kinds of aquatic arthropods, there are also many terrestrial species and numerous species of flying insects—the only invertebrates that have evolved a capability for flight. The body of arthropods is covered externally with a chitinous exoskeleton and, in varying degrees, is segmented both internally and externally. The jointed appendages are among the most apparent and versatile features of the arthropods and they have been adapted for walking, swimming, jumping, feeding, reproduction, sound production, sensory perception, and defense.

Among the important distinguishing characteristics of the Phylum Arthropoda are the following: (1) a hardened, chitinous exoskeleton; (2) paired, jointed appendages, often modified for various functions; (3) segmented body divided into two or three functional regions, such as head and trunk, cephalothorax and abdomen, prosoma and opisthosoma, or head, thorax, and abdomen; (4) reduced coelom; (5) complete digestive tract with mouthparts developed from modified anterior appendages; (6) an open venous system but a closed arterial system with a dorsal heart, arteries, and open spaces (sinuses) in the tissues that serve to collect blood prior to return to the heart; (7) a highly organized brain and a nervous system with paired ventral ganglia and nerve cords; and (8) complex sense organs and behavior.

Fig. 13.1 *Fossil trilobite, portion of head surface broken away.*

In this exercise, you will have an opportunity to study the organization of representatives of three of the four major groups (subphyla) of arthropods: the **chelicerates** (horseshoe crab, spider); the **crustaceans** (crayfish, water flea); and the **uniramians** (grasshopper, honeybee).

Classification

The classification of the arthropods is complicated because the phylum is so large and diverse. We can present only a brief summary of some of the major classes here. Recent studies have led to some changes in the classification of the phylum. Most current authorities recognize four major groups of arthropods—the trilobites, the chelicerates, the crustaceans, and the uniramians.

Subphylum Trilobita (Trilobitomorpha)

The trilobites (figure 13.1). Primitive marine arthropods with a flattened, ovoid body divided into three parts, an anterior head (cephalon), a middle thorax, and a posterior pygidium ("tail"). Two longitudinal grooves divide the body into three lobes. One pair of simple antennae on the head; all other appendages biramous and similar. All extinct.

Subphylum Chelicerata

The chelicerates. Body with distinct prosoma and opisthosoma, or prosoma and opisthosoma broadly fused. Antennae absent. Six pairs of appendages, including one pair of chelicerae (specialized pincerlike appendages modified for feeding), but no mandibles.

Class Merostomata (Water Scorpions and Horseshoe Crabs)

Marine animals with abdominal gills and one pair of lateral compound eyes. Examples: *Eurypterus* (water scorpion, fossil species only), *Limulus* (*Xiphosura*) (horseshoe crab).

Class Pycnogonida (Sea Spiders)

Small, marine, spiderlike chelicerates; well-developed cephalothorax and reduced abdomen. Example: *Pycnogonum*.

Class Arachnida (Scorpions, Spiders, Ticks, and Mites)

Mainly terrestrial animals, although a few inhabit aquatic areas, with six pairs of appendages. From the anterior, the first pair of appendages are chelicerae with claw or fang; the second pair are pedipalps (variously modified for sensory perception, sperm transfer, grasping); the remaining four pairs are walking legs. Head and thoracic areas fused to form cephalothorax (=prosoma). Prosoma and opisthosoma distinct and narrowly connected by pedicel, or else broadly fused. Examples: *Centruroides* (scorpion), *Argiope* (spider), *Dermacentor* (tick), *Trombicula* (mite).

Subphylum Crustacea

The Crustaceans. Mainly aquatic animals with cephalothorax (fused head and thorax) usually covered by a dorsal carapace; hardened chitinous exoskeleton; biramous appendages. Head with two pairs of antennae, one pair of mandibles, and two pairs of maxillae (accessory mouthparts).

Class Branchiopoda (Branchiopods)

Small crustaceans with trunk appendages flattened and leaflike, often modified for filter-feeding. Commonly live in temporary ponds or pools, mainly freshwater. Examples: *Daphnia* (water flea), *Artemia* (brine shrimp).

Class Cirripedia (Barnacles)

Adults sessile and modified for sessile or parasitic existence; carapace covered with calcareous plates; six pairs of thoracic appendages (cirri) modified for feeding. All marine. Examples: *Balanus* (acorn barnacle), *Lepas* (goose barnacle).

Class Malacostraca

Large crustaceans with trunk consisting of fourteen segments plus a posterior telson, thorax with eight segments, and abdomen with six segments. All segments typically bear appendages, first antennae, walking legs, and anterior abdominal appendages biramous; first one, two, or three pairs of thoracic appendages modified as maxillipeds. Examples: *Homarus* (lobster), *Penaeus* (shrimp), *Procambarus* (crayfish), *Callinectes* (blue crab).

Subphylum Uniramia (Uniramians)

Mainly terrestrial animals with uniramous (unbranched) appendages; head with one pair of antennae, one pair of mandibles, and one or two pairs of maxillae. Respiration via tracheal tubules and spiracles; excretion by Malpighian tubules.

Class Diplopoda (Millipedes)

Cylindrical body; one pair of short antennae; two pairs of walking legs per abdominal segment. Example: *Narceus* (*Spirobolus*).

Class Chilopoda (Centipedes)

Elongate, flattened body with many segments; one pair of antennae; abdominal segments with one pair of walking legs. Examples: *Lithobius, Scolopendra.*

Class Pauropoda (Pauropods)

Tiny cylindrical body; one pair of triramous (three-branched) antennae; no eyes; abdomen with nine or ten pairs of walking legs. Example: *Pauropus.*

Class Symphyla (Garden Centipede)

Mostly minute, slender, elongate bodies with fourteen segments; one pair of long antennae; one pair of spinnerets on last segment; twelve pairs of walking legs. No eyes. Example: *Scutigerella.*

Class Insecta (Insects)

Body with three regions: head, thorax, and abdomen; one pair of antennae; three pairs of walking legs and usually two pairs of wings. Examples: *Romalea* (grasshopper), *Musca* (housefly), *Apis* (honeybee), *Pulex* (flea).

Materials List

Living specimens
  Crayfish (*Procambarus*, etc.)
  Daphnia
  Honeybee (*Apis*)
  Spider (*Argiope* or similar)

Preserved specimens
  Limulus
  Romalea
  Argiope
  Crayfish, female with eggs (Demonstration)
  Representative crustaceans (Demonstration)
  Queen and drone bees (Demonstration)
  Honeybee, larva and pupa (Demonstration)

Prepared microscope slides
  Crayfish, compound eye (Demonstration)
  Crayfish, gill, cross section (Demonstration)
  Honeybee, mouthparts (Demonstration)
  Honeybee, spiracle (Demonstration)
  Insect cuticle (Demonstration)
  Argiope, walking leg

Plastic mounts
  Argiope
  Peripatus
  Lepas
  Balanus
  Crayfish appendages
  Millipede
  Centipede
  Honeybee
  Insect life cycle

Subphylum Chelicerata

The Horseshoe Crab: *Limulus*

The horseshoe crab, *Limulus* (Xiphosura) *polyphemus* (figure 13.2), is common in the shallow waters and sandy shores of the Atlantic coast of North America from Canada to Mexico. It is one of the few surviving members of an ancient group of chelicerate arthropods some 200 million years old.

■ Study a preserved or dried specimen of *Limulus* and observe the tough, leathery **carapace** that covers the **cephalothorax** (fused head and thorax) (figure 13.3).

Behind the horseshoe-shaped cephalothorax is a tapering, hexagonal **abdomen** and a long posterior **telson.** On the dorsal surface of the carapace, identify the two lateral

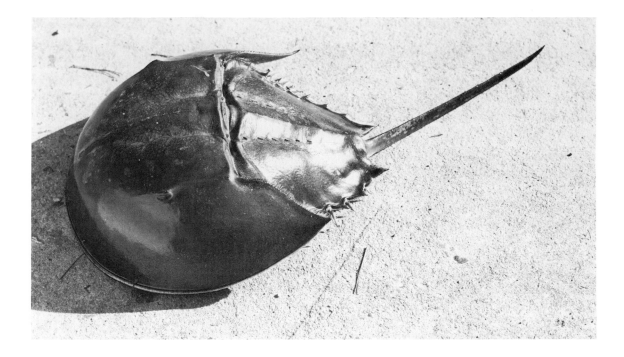

Fig. 13.2 *Horseshoe crab. (Courtesy of Carolina Biological Supply Company, Burlington, N.C.)*

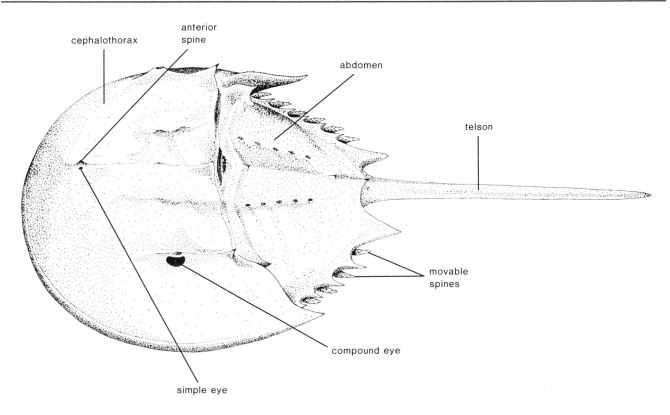

Fig. 13.3 *Horseshoe crab, dorsolateral view.*

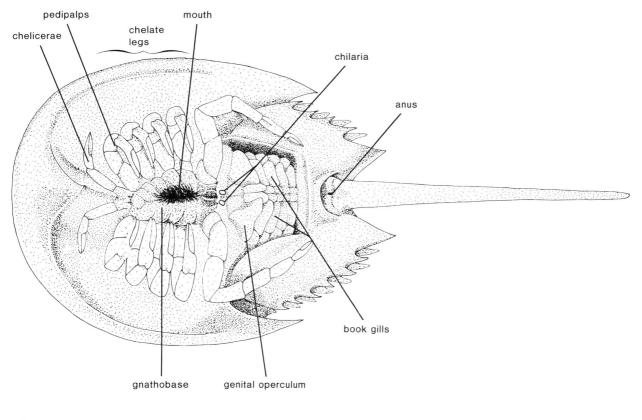

Fig. 13.4 *Horseshoe crab,* Limulus, *ventral view.*

compound eyes and the two **simple eyes** located on opposite sides of a small anterior spine. Along the sides of the abdomen find six pairs of **movable spines.**

On the ventral surface of the cephalothorax (figure 13.4), identify the **six pairs of appendages**: an anterior pair of **chelicerae** located in front of the mouth, four pairs of **chelate** (pincer) **legs,** a longer sixth pair of appendages with specialized tips for cleaning the gills and pushing the carapace through the sea bottom sediments. Behind the sixth pair of appendages, locate a single pair of small, spiny **chilaria.** On the abdomen find the six pairs of broad, flat structures, one pair per abdominal segment. The first pair of plates forms the **genital operculum,** which bears a pair of genital pores through which the gametes are shed. Behind the second to sixth pair of abdominal appendages are the many thin, leaflike, highly vascularized folds of the **book gills.** When submerged, these abdominal folds beat almost constantly to provide a continuous flow of seawater over the respiratory surfaces of the book gills.

The **mouth** is located ventrally near the center of the base of the walking legs. The spiny **gnathobases** at the base of the legs shred the food and push it toward the mouth.

Limulus feeds on worms, molluscs, and other small invertebrates in shallow ocean sediments. The crab plows through the sand by arching its back and pushing against the sand with the sixth pair of appendages and the telson.

Awkward as it may sound, this crude form of locomotion and feeding has been effective for some 200 million years. *Limulus* is also able to swim short distances by vigorous beating of the abdominal platelike appendages.

The Spider: *Argiope*

Spiders, belonging to the Order Araneida (=Araneae), are an ancient group, with some fossils dating back over 300 million years. With two exceptions, Antarctica and the oceans below the intertidal zone, spiders can be found throughout the world. Of an estimated world total of 50,000 species, about 35,000 have been named and described. Sometimes spiders are very abundant; grassy fields may contain over two million spiders per hectare. All spiders are carnivorous, and thus are important for controlling insect pests. The common garden spider serves well as an example of this important group of chelicerate arthropods (figure 13.5).

■ Obtain a preserved or plastic embedded specimen of the garden spider *Argiope* (formerly called *Miranda*) and study it under your dissecting microscope.

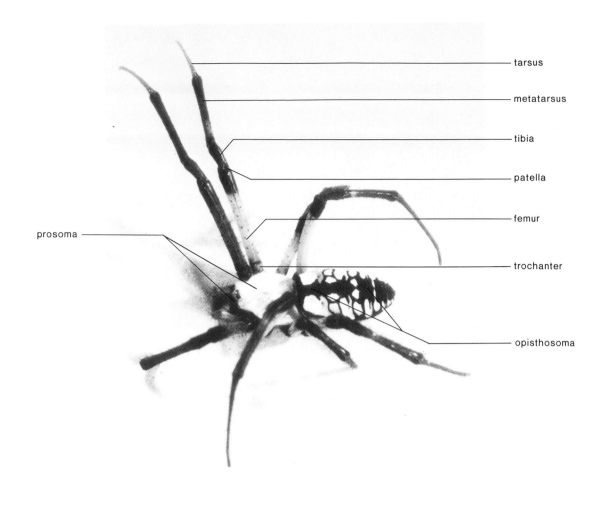

tarsus

metatarsus

tibia

patella

femur

trochanter

prosoma

opisthosoma

Fig. 13.5 *Garden spider,* Argiope. *(Courtesy of Carolina Biological Supply Company, Burlington, N.C.)*

The anterior region from which the legs project is called the **prosoma** (=cephalothorax) and is composed of the head and thorax fused together. The posterior part is called the **opisthosoma** (=abdomen). The two regions are connected by a narrow **pedicel.** The pedicel is seen most easily by gently bending the spider to expose the area between the overlapping opisthosoma and the prosoma. If your specimen is of a genus other than *Argiope,* the shape and size may vary, but the anatomical parts will be similar.

Observe that when spiders die, their legs close over the ventral side. The reason for this is that spiders do not have extensor muscles for their legs; they have only flexor muscles. Living spiders extend their legs by using blood as a hydraulic fluid.

Externally, the body of a spider is covered by a **chitinous exoskeleton,** which consists of numerous hardened plates called **sclerites** (for example, carapace, sternum). The sclerites are separated by sutures or thin membranous areas.

The body wall is made up of three main layers: an outer **cuticle;** an underlying **hypodermis,** which secretes the cuticle; and a thin, noncellular **basement membrane.**

Insects have a body wall with a similar structure. The sclerites serve a number of functions: body rigidity, muscle attachment sites, prevention from desiccation, and protection from predators.

Male and female spiders are usually very different in appearance from one another. The females may grow to be ten or more times larger than the males and may be very differently marked and colored. Such difference between the male and female is called **sexual dimorphism.**

The head region of the spider bears the eyes and mouthparts (figure 13.6). Most spiders have eight eyes (four pairs), but some have fewer or even none at all. Garden spiders have eight simple eyes called **ocelli,** visible at the anterior end of the carapace. The **carapace** is the portion of the exoskeleton covering the prosoma (fused head and thorax). Vision plays a relatively minor role in the behavior of most spiders; they rely more on tactile and chemical senses than vision to locate their prey. Eyes often aid in detecting subtle changes in light intensity and motion; but, with the exception of hunting spiders (such as the wolf, crab, and jumping spiders), the eyes of spiders do not form actual images.

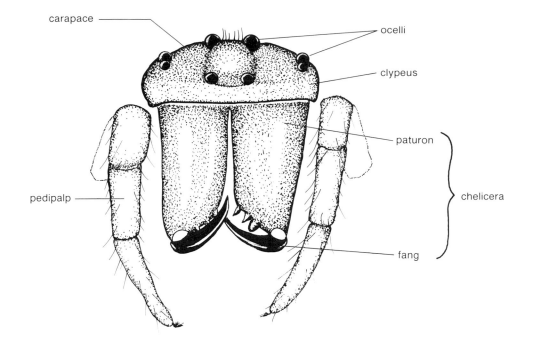

Fig. 13.6 *Garden spider,* Argiope, *anterior view of head.*

The area immediately below the eyes and down to the edge of the carapace is called the **clypeus.** Extending downward from below the clypeus is a pair of **chelicerae.** The chelicerae consist of a thick basal segment called the **paturon** and a small distal segment called the **fang** (figure 13.7). The fang articulates with the paturon. It is with the fang that a spider bites and poisons its prey.

Lateral to and slightly behind the chelicerae is a pair of **pedipalps** (=palps) that resemble small legs. (If your specimen is a male, the distal end of the pedipalp is enlarged to form the copulatory organ.) Now, turn the spider ventral side up and gently pry the legs away from the underside of the spider. Pedipalps consist of only six segments (figure 13.8a and b). The basal segment (coxa) of the pedipalp is expanded to form the **maxilla** or **endite** (=gnathobase). The endite often has a brush of hairs at its distal end called the **scopula.** The scopula aids in sponging up fluids from the prey as the spider feeds. The endites are manipulated to squeeze the fluids from the prey.

■ Examine a walking leg from your specimen and in a prepared microscope slide. Identify the parts as shown in figure 13.8c. Compare the segments of a pedipalp (figure 13.8a and b) with those of a walking leg. How do these appendages illustrate adaptation? Which segments of the pedipalps show the highest degree of specialization?

Fig. 13.7 *Fang of spider. (Scanning electron micrograph courtesy of North Carolina State University.)*

Arthropoda

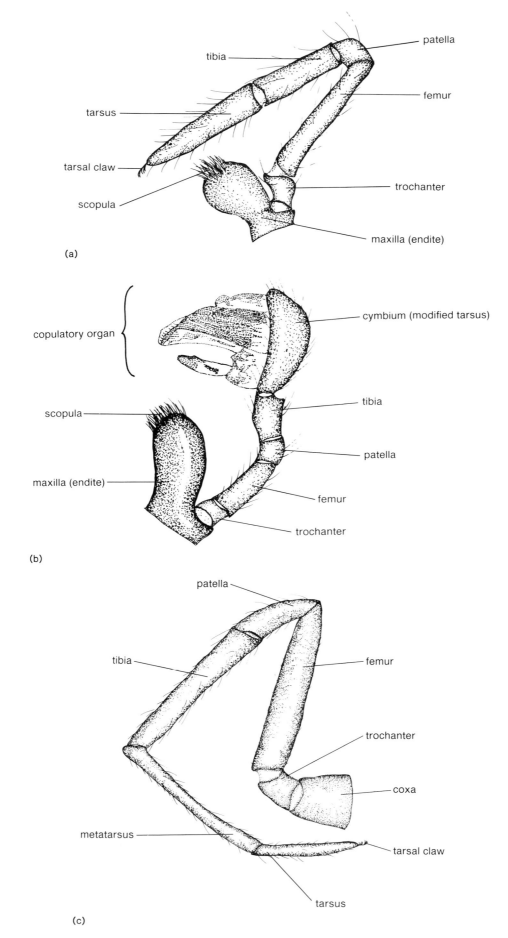

Fig. 13.8 *Garden spider*, Argiope. *(a) pedipalp of female; (b) pedipalp of male; (c) walking leg.*

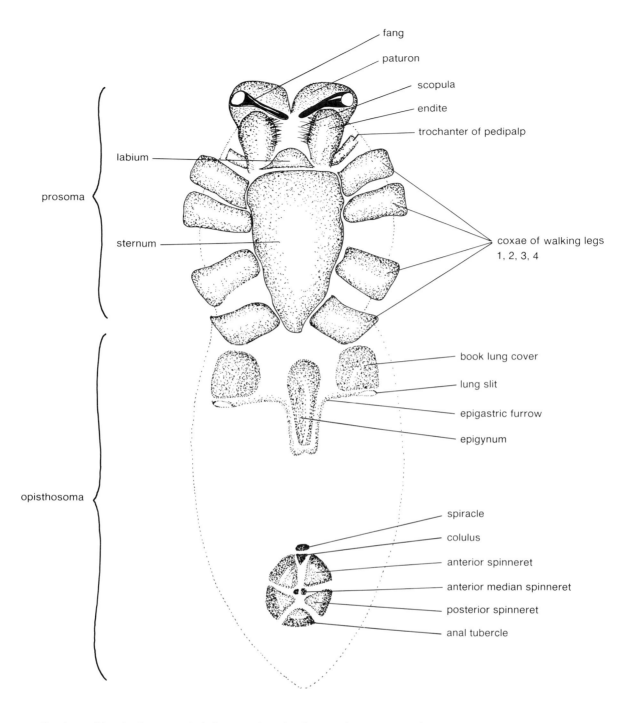

Fig. 13.9 *Garden spider,* Argiope, *ventral view, most parts of appendages removed.*

The **labium** (see figure 13.9) is located on the ventral surface between the two endites and immediately anterior to the sternum. The labium is sternite No. 2, or the ventral plate of the second fused segment.

The mouth (not visible) is located between the area formed by the endites, labium, and chelicerae. Note that there are no mandibles with which to chew the food. Spiders can eat only liquid foods. When spiders "eat" their prey, they first suck out the blood. Then, through the wound, they pour in digestive juices (the chief enzymes are proteases and lipases) that convert the prey's inner tissues to a soup. The soup is then sucked out by the spider, leaving the hollow remains of the prey.

The **sternum** is the large plate on the ventral surface of the prosoma (figure 13.9). It is surrounded laterally by the coxae of the four pairs of walking legs. The sternum represents the fused segments of sternites Nos. 3–6.

The pedicel connects the prosoma to the opisthosoma, and represents the seventh body segment. Except in the very primitive spiders (Suborder Liphistiomorpha), none of the original opisthosomal segmentation is evident in spiders. From histological studies, we know that the opisthosoma in spiders represents segments 7 to 17, and that the spinnerets come from segments 9 and 10, and the anus from segment 18.

On the ventral, anterior portion of the opisthosoma, find the distinct groove that runs from one side to the other (figure 13.9). This is the **epigastric furrow.** At each end of the epigastric furrow is a **lung slit,** the entrance to a **book lung.** Most of the air that larger spiders breathe enters through the lung slits. On the central, anterior edge of the epigastric furrow are the **gonopores.** If your specimen is a female, there will be a large, intricate, sclerotized plate extending posteriorly, the **epigynum.** If your specimen is a male, a small plate may or may not be visible. Male spiders do not have a penis. Instead, the palpal organ of the pedipalp is enlarged and serves as the **copulatory organ** (penis analogue) (figure 13.8b). Some workers believe that the copulatory organ may have evolved from the tarsal claw.

Near the ventral, posterior end of the opisthosoma, locate the three pairs of **spinnerets** (see figure 13.9). Two pairs, the anteriors and posteriors, are larger and distinct; the anterior median pair is smaller and tucked in between the larger pairs. The single **anal tubercle** is located at the central posterior edge of the spinneret cluster. Between the anterior bases of the anterior spinnerets find the small organ called the **colulus.** It is considered to be a small, vestigial spinneret. Immediately anterior to the spinnerets, in a groove, is an extremely small **spiracle,** which serves as the entrance to the tracheal system. As the spiracle is all but invisible, it is not suggested that you look for it.

Turn the spider over and observe the dorsal side. On the central, slightly posterior part of the carapace, is a depression that represents the **cervical groove** (see figure 13.10). Aside from marking the boundary between the cephalic and thoracic regions, it is also an apodeme, serving as a muscle attachment site on the inside of the carapace.

Manipulate the spider gently so that the pedicel can be seen from above. On the dorsal surface of the pedicel is a small pair of plates called the **lorum** (may be difficult to see).

There are no external, dorsal features worthy of note on the opisthosoma of spiders. However, in the central, dorsal anterior region is the cardiac area, beneath which lies the longitudinal **heart** that pumps blood anteriorly. Pale markings on the opisthosoma result from the deposition of guanine, an excretory product of spiders.

Demonstrations

1. Living *Argiope* or other spider in terrarium.
2. Plastic mount of *Argiope.*

Subphylum Crustacea, Class Branchiopoda

A Water Flea: *Daphnia*

The water flea, *Daphnia* (figure 13.11), is a common microscopic crustacean found in many bodies of fresh water. Its structure is simpler than that of the crayfish, and because of its transparency and small size, living *Daphnia* can be studied easily in the laboratory. *Daphnia* and other members of the Class Cladocera of the Subphylum Crustacea are popularly known as "water fleas" because of their characteristic swimming movements. These animals constitute an important element in the diet of many fish and other, larger, aquatic animals.

■ Make a wet mount of a living *Daphnia* on a microscope slide and observe its characteristic shape.

External Anatomy

Daphnia swims by rapid movements of its two large **antennae** and feeds on microscopic food particles filtered from the water by means of complex movements of five pairs of thoracic appendages, which bear many setae. Food particles removed from the water are collected in a **median ventral groove** located at the base of the legs and are passed forward to the mouthparts where larger particles are macerated by the sclerotized (hardened) **mandibles** before passing into the mouth.

The body of *Daphnia* is laterally compressed and not obviously segmented. It consists of three main regions— an anterior head, a large thorax, and a smaller postabdomen. A true abdomen is absent in the Cladocera. The thorax is covered by a part of the chitinous exoskeleton called the **carapace,** which has a bivalved appearance, but which is fused dorsally and open ventrally. The compact head is fully enclosed in the exoskeleton (that is, it does not open ventrally) and is bent downward. Under certain conditions, some Cladocera exhibit seasonal changes in the morphology of the head by forming a dorsal elongation or "helmet." (Certain other parts of the body may be modified also.) This kind of morphological variation is called **cyclomorphosis,** and there is evidence that it is influenced by temperature, turbulence of the water, heredity, and other factors.

The most conspicuous structure on the head is the single large compound eye with a central mass of pigmented granules surrounded by numerous hyaline lenses. A single small *ocellus* (simple eye) lies posterior to the compound eye.

Locate the first and second antennae. The first antennae, or antennules, are small and unsegmented, and are located on the ventral side of the head near the posterior margin. They bear chemical sense receptors (specialized setae) and are primarily sensory in function. The tapering projection of the head between the antennules is the **rostrum.**

The second antennae are very large structures attached laterally near the posterior margin of the head. Each second antenna consists of a stout, unjointed **basal segment,** and segmented **dorsal** and **ventral rami.** Each ramus consists of several segments and bears numerous **plumose** (featherlike) **setae.** The second antennae are moved by a set of strong muscles that originate on the dorsal side of the carapace near the junction of the head and the thorax. Ventrally, near the same junction, the mouth may be found. It is surrounded by six mouthparts:

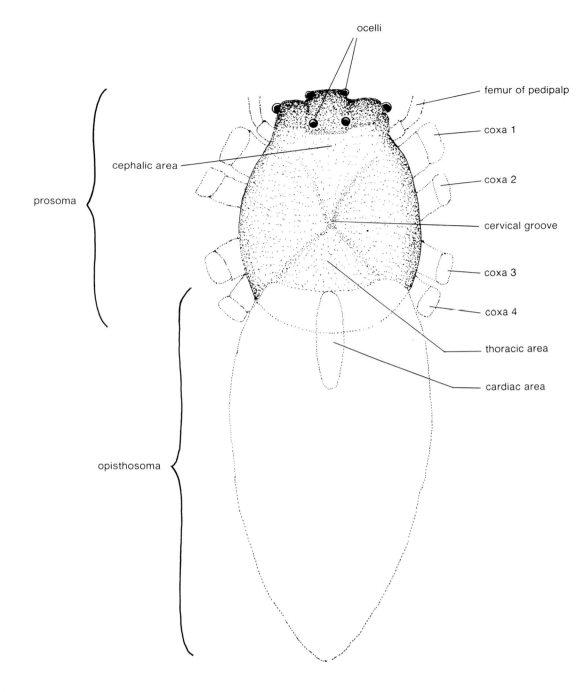

Fig. 13.10 *Garden spider,* Argiope, *dorsal view, most parts of appendages removed.*

(1) a dorsal median **labrum** (upper lip); (2) a pair of stout **mandibles;** (3) two small pointed **maxillae,** which are used to push food between the mandibles; and (4) a ventral median **labium** (lower lip).

The **thorax** consists of six segments, and all but the fifth segment bear paired appendages. These five pairs of lobed, leaflike (foliaceous) appendages are basically biramous, but this condition is masked by their high degree of specialization. All of the thoracic appendages bear many hairs and setae used to strain food particles from the water.

As previously stated, a true abdomen is absent in *Daphnia,* but there is a definite **postabdomen** located at the posterior end of the body. It is usually held bent forward under the thorax so that the dorsal side of the postabdomen is directed downward. The postabdomen bears no limbs; but it has two rows of spines, and it terminates in two long abdominal setae. It is equipped with an effective musculature and appears to function mainly in cleaning debris from the thoracic legs, although it may also aid locomotion.

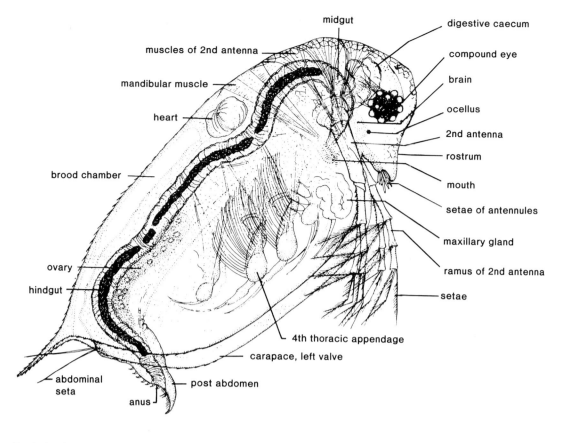

Fig. 13.11 Daphnia, *lateral view.*

Internal Anatomy

The complicated musculature of *Daphnia* tends to obscure some of the smaller anatomical details, but the major elements of most of the internal systems can be seen in a whole specimen. Study as many as possible of the following features on your specimen.

1. **Digestive system**—Relatively unspecialized, and consisting of the **mouth, foregut** (esophagus) with a cuticular lining, **midgut** (stomach-intestine), and a **hindgut** (rectum), the latter also with a cuticular lining. Digestion takes place principally in the midgut region. Located anteriorly (in the head region above the compound eyes) are two digestive caeca, which probably secrete digestive enzymes.

2. **Circulatory system**—A simple oval **heart** lies behind the head and dorsal to the gut. Blood enters through two lateral **ostia** and leaves via an anterior opening. There are no definite blood vessels, and the blood is roughly channeled through the large central **haemocoel** by a series of thin mesenteries.

3. **Respiratory system**—None. Exchange of gases occurs through the body wall, especially through that portion lining the inner surfaces of the valves, and it occurs, also, through the surfaces of the legs.

4. **Excretory system**—Two looped **maxillary glands** located near the anterior end of the valves are thought to be excretory in function.

5. **Nervous system**—Double **ventral nerve cord,** a few **ganglia,** paired **nerves,** and a **brain** located between the foregut and the compound eyes.

6. **Reproductive system**—Two elongated **ovaries** lie lateral to the midgut in the female *Daphnia*. Posteriorly, each ovary is connected to the **brood chamber** (a cavity dorsal to the body proper and located between the valves of the carapace; closed posteriorly by dorsal processes of the postabdomen) by a thin **oviduct.**

In the male, two **testes** occupy the same location but are smaller in size and are continued posteriorly as **sperm ducts,** which follow the gut and open on the postabdomen near the anus. In some species of Cladocera, a part of the postabdomen of the male is modified to form a copulatory organ.

Under favorable conditions, a female *Daphnia* lays thin-shelled, diploid eggs, which contain little yolk and develop parthenogenetically. Under adverse conditions, however, the females produce thick-shelled, resistant eggs with large yolk reserves. The latter type of eggs are haploid and require fertilization by a male for their development. Both types of eggs are deposited in the brood chamber.

Male *Daphnia* appear in natural populations only at certain times of the year and often constitute a small proportion of the total population. The specific factors responsible for the production of males are not known, despite much investigation, but it appears that male production can be induced by a combination of environmental factors: crowding, decreasing food supply, and water temperature. These environmental factors appear to affect the metabolism of the female in such a way that parthenogenetic male eggs rather than parthenogenetic female eggs are released into the brood chamber.

There are no larval stages in the development of the Cladocera (a situation not at all typical of Crustacea!), but four different periods can be recognized in the life history of a cladoceran: egg, juvenile, adolescent, and adult.

Parthenogenetic eggs develop in the brood chamber into a stage already having the adult form before their release (first instar young). Subsequent development continues through a series of molts in which the carapace and other chitinous portions are lost and the animal increases in size. The average life span is said to cover about seventeen instars (instar = intermolt stage).

Resistant eggs are retained in the brood chamber until the next molt. A portion of the brood chamber surrounding them is modified into a thick-walled protective case called an **ephippium.** At the next molt of the parent, the ephippium and its contained eggs are shed.

Subphylum Crustacea, Class Malacostraca

The Crayfish: *Procambarus*

The crayfish is a large aquatic arthropod, which effectively illustrates many of the basic characteristics of the phylum. Large southern crayfish (*Procambarus*) or western crayfish (*Astacus*) are especially suitable for laboratory study because of their large size, but members of other common genera, such as *Cambarus* and *Orconectes,* can also be used. The American lobster *Homarus* is very similar in structure to the crayfish and may also be used when available.

External Anatomy

■ Obtain an anesthetized or freshly killed crayfish and study the general organization of the body (figures 13.12 and 13.13). Observe that the body is divided into two major regions: an anterior **cephalothorax,** which is made up of a fused head and thorax, and a posterior **abdomen.**

Extending anteriorly from the cephalothorax between the compound eyes is a pointed **rostrum.** Find the **cervical groove,** which represents the line of fusion between the head and the thorax. The hard outer covering (exoskeleton) of the cephalothorax is the carapace; note that the carapace and the chitinous exoskeleton on the dorsal side of the abdomen are hardened by the deposition of mineral salts (except in newly molted specimens). The crayfish body is made up of a total of nineteen segments. Segments I–XIII make up the cephalothorax (five in the head and eight in the thorax), and segments XIV–XIX constitute the abdomen.

Observe the two large **compound eyes** on movable eyestalks, two large **antennae** (sometimes called second antennae), and two smaller **antennules** (sometimes called first antennae). The pointed rostrum projects forward between the eyes. The **mouth** is ventral, surrounded and largely concealed by several pairs of modified appendages that serve as mouthparts, and will be studied later. Locate the five pairs of large **walking legs** and five pairs of small abdominal appendages or **swimmerets.** The first pair of walking legs bear large pincers or chelae and thus they are called **chelipeds** ("pincer legs"). How many of the other pairs of walking legs also bear chelae? On the last abdominal segment is a pair of large flattened lateral appendages, the **uropods.** Find the **anus** on the ventral side of the **telson,** a medial extension of the last abdominal segment (not an appendage).

Crayfish are dioecious, and each sex exhibits distinctive secondary sexual characteristics. Females have a broad abdomen, rudimentary appendages (swimmerets) on the anterior abdominal segments, a **seminal receptacle** on the ventral surface of the exoskeleton between the fourth and fifth pairs of walking legs, and two female sex openings located at the bases of the third pair of walking legs. Males have a narrower abdomen, enlarged swimmerets (**gonopods**) on the first abdominal segments (modified for the transfer of sperm to the seminal receptacle of the female), and openings of the vasa deferentia (male sex ducts) at the base of each fifth walking leg.

■ Determine the sex of the specimen you are examining and observe also specimens of the opposite sex distributed around the laboratory.

Appendages

In early ancestors of the arthropods, the appendages of all body segments were biramous and similar on all segments. During the evolution of arthropods, various segments and the attached appendages have become adapted to serve in sensory perception, defense, swimming, walking, grasping, sperm transfer, and so forth. Since the body segments and appendages arise similarly during embryonic development, they are said to be homologous structures.

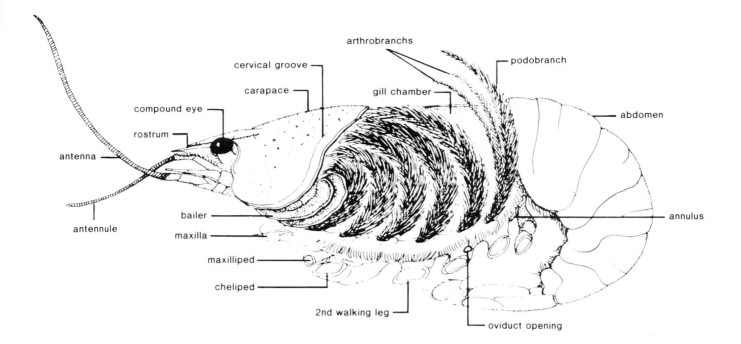

Fig. 13.12 *Crayfish, lateral view, portion of carapace removed to show gills.*

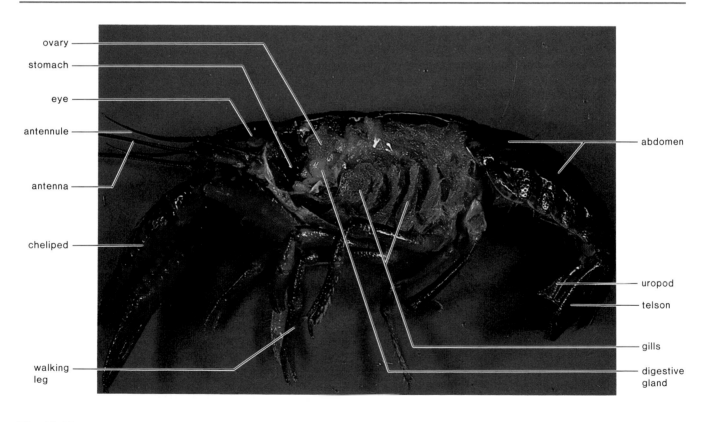

Fig. 13.13 *Crayfish, lateral view, part of carapace removed. (Photograph by Ken Taylor.)*

Homology is an important principle in zoology, and the study of structural homologies has provided much valuable evidence about the evolutionary history and relationships of animals.

The paired appendages of the crayfish provide an excellent illustration of **serial homology,** the adaptation of a longitudinal series of originally similar organs to carry on different functions. The appendages of primitive arthropods were **biramous** (two-branched). This is the simplest type of arthropod appendage and is most nearly approximated in the crayfish by the swimmerets of the abdominal segments.

■ Remove an appendage from the third abdominal segment by cutting it free near its attachment, and observe that it consists of a basal portion, the **protopodite,** and two terminal branches, a lateral (outer) **exopodite** and a medial (inner) **endopodite.**

Carefully remove the appendages from one side of the body, starting with the first walking leg (cheliped) and working forward and then backward. Remove each appendage close to its attachment to the body and arrange it in order on a sheet of paper as in figure 13.14. Study each appendage with the aid of figures 13.14 and 13.15 and the descriptions provided following.

1. **Antennule** (first antenna) base of three segments, and two long, many-jointed filaments. This appendage is probably not truly biramous. The slender inner branch arises from the base of the outer thicker filament and is, therefore, probably not homologous with an endopodite. The two basal segments are homologous with the protopodite. Within the basal segment of each antennule is a **statocyst,** a saclike sense organ that opens dorsally via a small pore.

2. **Antenna** (second antenna) protopodite of the two segments; endopodite long, many-jointed; exopodite a short scale.

3. **Mandible** basal segment of protopodite greatly enlarged as functional jaw; second segment small and forming the base of the palpus; endopodite, the two small distal segments of the palpus; exopodite lost.

Note: The metastoma, or bifurcated "lower lip" between the mandible and the first maxilla, is a leaflike blade that fits closely over the convex surface of the mandible. It is not a true appendage.

4. **First maxilla** protopodite of two flat, leaflike segments; endopodite of two segments, the distal one narrow and pointed; exopodite lost.

5. **Second maxilla** protopodite of two flat, bilobed segments; endopodite, a small pointed segment; exopodite forming part of a large elongated plate, the **bailer.**

Fig. 13.14 *Crayfish appendages. (Photograph courtesy of Indiana University Audiovisual Center.)*

6. **First maxilliped** protopodite of two flat segments extending inward; endopodite and exopodite are both present.

7. **Second and third maxillipeds** protopodite of two segments; endopodite of five distinct segments; exopodite relatively thin.

8. Five walking legs or **pereiopods** protopodite of two segments; endopodite of five joints as in the second and third maxillipeds; exopodite lost. In the large chelipeds (first walking legs), the second segment of the protopodite and the first segments of the endopodite are fused together.

9. **First and second swimmerets** the first is rudimentary in the female; in the male, the protopodite and the endopodite are fused, and the entire structure is modified for transferring spermatozoa to the seminal receptacle of the female; exopodite is lost.

10. **Third to fifth** (second to fifth in the female) **abdominal appendages** (swimmerets) protopodite of two segments; endopodite and exopodite flat and filamentous.

	appendage	function
head	antennule	equilibrium, touch, taste
	antenna	touch, taste
	mandible	biting food
	1st maxilla	} food handling
	2nd maxilla bailer	
thorax	1st maxilliped	} touch, taste, food handling
	2nd maxilliped	
	3rd maxilliped	
	cheliped	offense, defense
	2nd "walking leg"	} walking, grasping
	3rd "walking leg"	
	4th "walking leg" gills	} walking
	5th "walking leg"	
abdomen	1st swimmeret female male	} in male, reproduction
	2nd swimmeret	
	3rd swimmeret	} H$_2$O circulation, carry eggs and young
	4th swimmeret	
	5th swimmeret	
	uropod	swimming, egg protection

☐ protopodite   ▨ endopodite   ■ exopodite

Fig. 13.15 *Functions of crayfish appendages.*

11. **Sixth abdominal appendage** (uropod) greatly enlarged and modified to form (with the telson) the tail fan, which is used for swimming backward.

Respiratory System

■ Locate the gills within the **gill chambers** at each side of the carapace (figures 13.12 and 13.13). The lateral flaps of the carapace, covering the gills, are termed **branchiostegites.** Remove the left branchiostegite carefully by cutting away the carapace with your scissors to expose the gills; make your cut carefully to avoid damage to the underlying gills.

The gills are feathery projections of the body wall, which contain blood channels. Water is pumped through the gill chamber by the action of the **bailer,** a paddlelike projection of one of the oral appendages (the second maxilla). Oxygen dissolved in the water diffuses across the thin walls of the gills and combines with the blood in the gills. Carbon dioxide diffuses across the walls of the gills in the opposite direction.

Observe that the gills occur in distinct longitudinal rows. How many rows of gills are there in the specimen? The outer row is attached to the base of certain of the appendages. Which ones? These outer gills are the **podobranchs** ("foot gills"). How many podobranchs do you find in the specimen? Separate the gills carefully with your probe or dissecting needle, and locate the inner row(s) of gills. These inner gills are the **arthrobranchs** ("joint gills") and are attached to the chitinous membrane, which joins the appendages to the thorax (see figure 13.12). How many rows of arthrobranchs do you find in the specimen?

■ Remove a gill with your scissors by cutting it free near its point of attachment and place it in a watch glass filled with water. Observe the numerous gill filaments arranged along a central axis. Study also a demonstration slide of a crayfish gill and note the afferent and efferent blood vessels in the central axis.

Internal Anatomy

■ Remove the remainder of the carapace up to the rostrum by cutting forward from the rear of the carapace up to a level just behind the eyes. Leave the eyes, rostrum, and adjacent anterior portion of the carapace intact. With your scalpel, carefully separate the hard carapace from the thin, soft underlying layer of **hypodermis** as you cut away the carapace. The hypodermis is the tissue that secretes the carapace and other parts of the exoskeleton. Next, remove the remaining gills on the left side by cutting them near their points of attachment, and cut away the chitinous membrane underlying the gills to expose the internal organs as in figure 13.13.

Circulatory System. Locate the small membranous **heart** just posterior to the stomach (see figure 13.16). If you are dissecting a preserved specimen, you may find the heart filled with colored latex.

Identify the **pericardial sinus** (the space in which the heart lies), the **ostia** (openings in the wall of the heart), and the delicate **arteries** attached to the heart. Because of their small size, the arteries are often difficult to locate. Carefully search among the internal organs and try to locate the following arteries and determine which structures they supply with blood: (1) **ophthalmic artery,** (2) two **antennary arteries,** (3) two **hepatic arteries,** (4) **dorsal abdominal artery,** and (5) **sternal artery.** The circulatory system of the crayfish is an open system; blood is pumped from the heart through the major arteries that branch into smaller arteries and supply blood to the various organs of the body. From the arteries, the blood flows into open spaces, or **sinuses,** within and between the organs and then into a large **sternal sinus** along the floor of the thorax. Channels from the sternal sinus lead to the gills where the blood is oxygenated before returning to the pericardial sinus and into the heart via the ostia.

Crayfish blood consists of a nearly **colorless plasma** containing a dissolved respiratory pigment **hemocyanin** (a complex copper-containing protein) and numerous amoeboid blood cells.

Muscular System. Locate first some of the principal muscles of the crayfish. These include the two longitudinal bands of **abdominal extensor muscles** that run along the dorsal side of the thorax and extend back along the dorsal surface of the abdomen (figure 13.17). What is their function? Find the large **abdominal flexor muscles** in the abdomen lying below the extensor muscles. These large muscles serve to bend or flex the abdomen; hence, they provide the force for the quick backward thrust of the animal when it is alarmed.

The **gastric muscles** attach the stomach to the inner wall of the carapace, and two large **mandibular** muscles **originate** (fixed end of the muscle) on the carapace and **insert** (movable end of the muscle) on the mandibles. Contraction of these muscles causes the crushing or grinding action of the mandibles. Each of the other appendages is also equipped with a musculature; **flexor muscles** serve to draw the appendage closer to the body or to its point of attachment, and **extensor muscles** serve to extend, or straighten out, the appendage.

Female Reproductive System. Locate the paired **ovaries** lying ventral and lateral to the pericardial sinus (figure 13.17) and trace the delicate **oviducts** to their openings at the base of the third walking legs. Are there **eggs** in the ovaries? What color are they? Locate again the **seminal receptacle** where sperm is deposited during copulation. Observe the demonstration of a female "in berry" with clusters of eggs attached to the bristles on the swimmerets.

Male Reproductive System. Find the paired **testes** beneath the pericardial cavity, and trace the **vasa deferentia** to their openings. Where are the male openings located? Compare the first abdominal appendages of the male specimen to those of a female specimen.

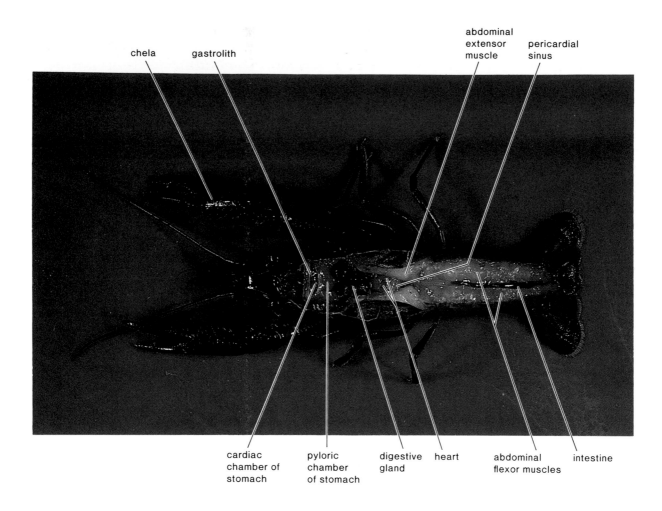

chela   gastrolith

abdominal
extensor   pericardial
muscle   sinus

cardiac   pyloric   digestive   heart   abdominal   intestine
chamber of   chamber   gland   flexor muscles
stomach   of stomach

Fig. 13.16   *Crayfish. Dorsal view, upper portion of exoskeleton removed. (Photograph by Ken Taylor.)*

Digestive System. The **mouth** is obscured by the several oral appendages and can be located by moving these appendages aside with your forceps. Locate, behind the mouth, the tubular **esophagus** leading to the **stomach** just behind the rostrum (figures 13.16, 13.17, and 13.18). The stomach is divided into a large anterior **cardiac chamber** and a smaller posterior **pyloric chamber.**

■ Cut open the stomach and study the **gastric mill,** a grinding apparatus consisting of three chitinous teeth. Observe in the pyloric chamber the **chitinous bristles,** which serve as a strainer to prevent large chunks of food from passing into the intestine. You may also find calcareous **gastroliths** attached to the walls of the stomach.

These calcareous bodies serve as stores of calcium salts, and they appear to play an important role in the calcification of the exoskeleton after molting. Locate the large digestive gland, or **hepatopancreas.** What is its function? Trace the **intestine,** behind the **stomach,** into the abdomen. Find the blind, saclike **intestinal caecum** in the abdomen, and the terminal **anus.**

Excretory System. The excretory structures of the crayfish are two large **green glands** (also called antennal glands) located ventrally in the head region near the base of the antennae. These glands are usually dark red or brown (not green!) in preserved specimens.

■ Find the two **excretory pores** at the bases of the antennae through which the wastes are discharged to the exterior.

Nervous System and Sense Organs

■ Carefully remove the organs of the reproductive, digestive, and circulatory systems in the head and thoracic regions to expose the ventral surface of the cephalothorax. Remove also the muscles and intestine from the abdominal region, carefully cut away the skeletal plates that cover the anterior part of the nervous system, and locate the **ventral nerve cord** (figure 13.19). Observe the **segmental ganglia** and their **paired lateral nerves.** Trace the nerve cord forward to the thorax and carefully cut away the skeleton that surrounds the nerve cord in this region.

Fig. 13.17 *Crayfish, internal anatomy, female.*

antennule

antennule nerve

antennary nerve

rostrum

green gland

antenna

cardiac stomach

mouth

esophagus

eye

brain

ophthalmic nerve

ophthalmic artery

antennary artery

ostia

heart

pericardial sinus

ovary

intestine

abdominal flexor muscles

abdominal extensor muscles

dorsal abdominal artery (divides)

intestine

anus

telson

uropod

swimmerets

ventral abdominal artery

ventral nerve cord

digestive gland

oviduct

opening of oviduct

sternal artery

sternal sinus

segmental ganglion

ventral thoracic artery

subesophageal ganglion

circumesophageal connective

hepatic artery

Arthropoda

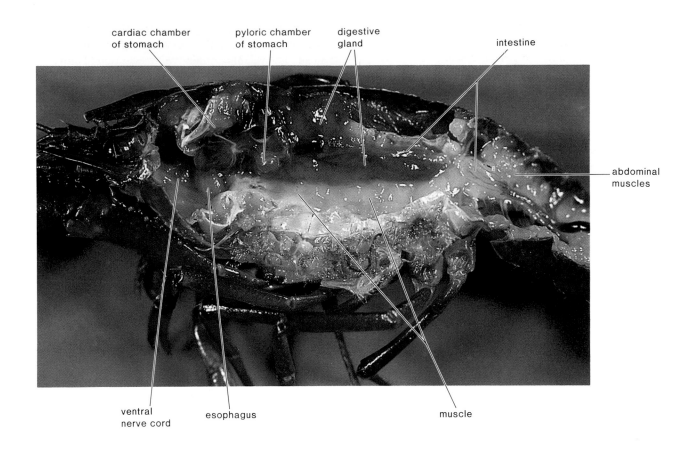

Fig. 13.18   *Crayfish, lateral view, gills removed to show internal organs. (Photograph by Ken Taylor.)*

Fig. 13.19   *Crayfish, internal organs removed to show ventral nerve cord. (Photograph by Ken Taylor.)*

Follow the nerve cord anteriorly and locate the **supraesophageal ganglion** (the "brain"), the **subesophageal ganglion,** and the **circumesophageal connectives.** Find the several pairs of nerves leading from the brain to the eyes, antennules, antennae, and mouthparts.

The most prominent sense organs of the crayfish are the **compound eyes** and the **statocysts,** both located previously. The crayfish also has numerous **sensory hairs** found on various appendages that are sensitive to touch. Special **chemoreceptors** on the antennae, antennules, and mouthparts also provide senses of taste and smell.

The statocysts are a pair of small sacs located in the basal segments of the antennules. They have a cuticular lining bearing sensory hairs. A sand grain within the sac, the **statolith,** stimulates the sensory hairs and thus provides a sense of equilibrium. The lining of the statocysts and the statoliths are lost and replaced at each molt.

Demonstrations

1. Compound eye (microscope slide).
2. Section of crayfish gill (microscope slide).
3. Female crayfish with eggs attached to swimmerets.
4. Live crayfish in aquaria for study of behavior.
5. Preserved specimens of several other types of freshwater and marine Crustacea.

The large compound eyes are made up of many individual units, the **ommatidia.**

■ Remove one of the compound eyes and examine its surface under your stereoscopic microscope. The surface of the eye is covered with a transparent **cornea** secreted by underlying cells. Observe that the cornea is divided into many sections, or **facets.**

Each facet of the cornea represents the outer end of an ommatidium. The compound eye forms a mosaic impression of the surroundings; also because of its structure, the compound eye is well adapted for detecting movements.

Subphylum Uniramia, Class Insecta

A Grasshopper: *Romalea*

Insects constitute the largest, most diverse, and most widespread class of arthropods, and indeed of all the multicellular animals. More than 750,000 species of insects have been described, and several thousand new species are added to the list each year. The insects are easily distinguished from the other classes of arthropods by a combination of the following characteristics of adult insects: (1) body divided into **three distinct regions** (head, thorax, and abdomen); (2) **one pair of antennae;** and (3) **three pairs of legs** and, usually, two pairs of wings.

Although the insects are an exceedingly diverse group of running, jumping, flying, swimming, and creeping organisms, the fundamental organization of the insect body is effectively illustrated by the grasshopper. Grasshoppers are members of a relatively unspecialized order of insects (Order Orthoptera), along with the katydids, locusts, and crickets. Specimens of the large, black, lubber grasshopper, *Romalea microptera* (figure 13.20), are usually provided for laboratory study, but the following information is generally applicable, also, to other species of grasshoppers.

External Anatomy

■ Obtain a preserved grasshopper and study its external features.

Externally, the body of a grasshopper is covered by a **chitinous exoskeleton** consisting of numerous hardened plates, or **sclerites,** separated by sutures, or thin membranous areas. The latter provide the flexibility necessary for movements of the body segments and the appendages. The body wall is made up of three principal layers—an outer **cuticle;** an underlying **hypodermis,** which secretes the cuticle; and a thin, noncellular **basement membrane.** The cuticle is a complex structure consisting of three layers: a very thin outer epicuticle, an exocuticle, and an innermost endocuticle lying directly upon the hypodermis. The endocuticle and exocuticle are composed largely of chitin, a nitrogenous polysaccharide, while the epicuticle is nonchitinous and consists chiefly of a network of fibrous proteins impregnated with wax. Study the demonstration provided in the laboratory, showing the microscopic structure of insect cuticle.

The head of the grasshopper (figure 13.21) bears the eyes, antennae, and mouthparts. Two large **compound eyes** occupy prominent sites on opposite sides of the head, and three **ocelli** (simple eyes) form an inverted triangle on the front of the head between the compound eyes. The two jointed **antennae** arising anterior to the compound eyes serve as organs of touch, smell, and taste. The **mouthparts** of a grasshopper are relatively unspecialized in comparison with those of most other kinds of insects and are of the **chewing type.** They consist of seven parts: a dorsal **labrum** or upper lip; a pair of sclerotized (hardened) **mandibles;** a pair of **maxillae,** which manipulate the food; a broad median **labium** or lower lip; and, in the center of these mouthparts, a cylindrical **hypopharynx** or tongue at the base of which the **salivary glands** empty. The labium and the maxillae bear sensory palps, which project ventrally. The mouthparts can be most easily observed and

Female

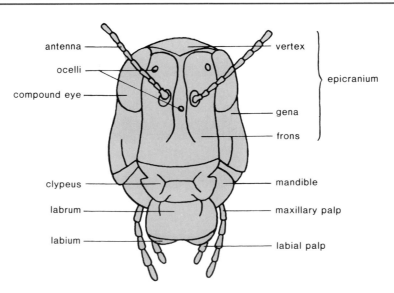

Male

Fig. 13.20 *Grasshopper,* Romalea, *lateral view of female and posterior view of male abdomen.*

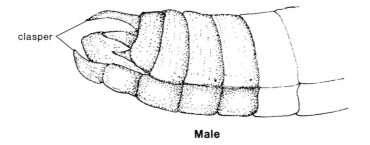

Fig. 13.21 *Grasshopper, head, anterior view.*

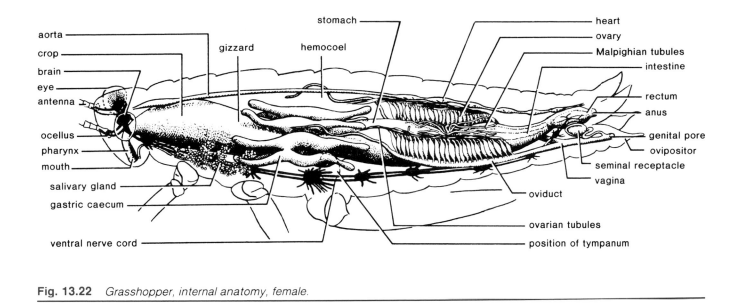

Fig. 13.22 *Grasshopper, internal anatomy, female.*

studied by removing them carefully from the head with your forceps and studying them under your stereoscopic microscope.

The thorax, like the head and abdomen, consists of several fused segments, illustrating the multisegmented ancestry of the insects. In the thorax, there are three segments, a large anterior **prothorax,** followed by two smaller segments, the **mesothorax** and the **metathorax.** Each segment bears one pair of legs. Also, the mesothorax and the metathorax each bear a pair of wings.

The legs are clearly jointed and consist of five main segments: a short **coxa,** which is attached to the ventral body wall; a small **trochanter** fused with a stout **femur;** a slender, spiny **tibia;** and a distal **tarsus** bearing two terminal claws. Between the claws is a fleshy pad, the **arolium,** which provides a grip on smooth surfaces. All three pairs of legs are used by the grasshopper for walking and climbing, but the third pair (the metathoracic legs) are specially modified for leaping. They are equipped with an enlarged femur containing strong voluntary muscles and with an elongated tibia to provide additional leverage. Another modification of the metathoracic legs can be seen in male specimens of *Romalea* and its close relatives. The inner surface of the femur bears a row of small spines and is rubbed against the lower edge of the front wing to produce the sounds that comprise the characteristic song of the males. Certain other species of grasshoppers produce their song by rubbing together their fore and hind wings. This type of sound production is called **stridulation.**

The long, narrow **fore wings** are borne on the mesothorax and serve as a covering for the hind wings when at rest. Observe the numerous **wing veins** in the thin, membranous **hind wings.** Note the difference in texture and pigmentation between the fore and the hind wings. The grasshopper wing develops as a saclike outgrowth of the body wall; this outgrowth later flattens to form a thin double membrane, and the two opposing membranes fuse together enclosing tracheae, nerves, and blood sinuses. The hollow veins are formed by thickening of the cuticle around the **tracheae** (air tubes) or along the sinuses. The pattern of wing venation in insects is very precise and characteristic, and therefore plays an important role in classification.

The cylindrical abdomen of *Romalea* consists of eleven segments, the last three of which are reduced in size and modified either for copulation or for egg laying (oviposition). Along the sides of certain of the thoracic and abdominal segments are ten pairs of tiny respiratory openings, the **spiracles,** which open into the system of **tracheal tubules.** The first abdominal segment also bears on each side a large, oval **tympanic membrane,** which serves as an organ of hearing. There is one pair of **sensory cerci** on the eleventh segment.

Internal Anatomy

Freshly killed or anaesthetized specimens are most suitable for the study of internal anatomy, although well-preserved specimens can also be used with some success. The internal tissues and organs are difficult to preserve, however, because of the chitinous exoskeleton of the grasshopper, which limits penetration of the preservative fluids. Study figure 13.22 for orientation before starting your dissection.

■ Remove the wings, and starting at the posterior end of the body, cut through the body wall along each side just above the spiracles. Remove the freed dorsal wall and pin back the sides of the remaining body wall in your dissecting pan. Cover your dissection with water to keep the internal organs moist and pliable.

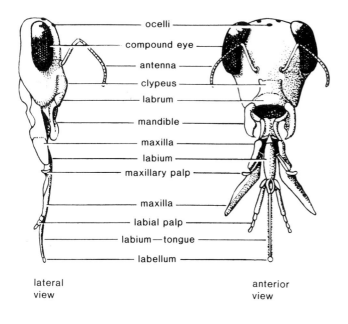

ocelli
compound eye
antenna
clypeus
labrum
mandible
maxilla
labium
maxillary palp
maxilla
labial palp
labium—tongue
labellum

lateral
view

anterior
view

Fig. 13.23 *Honeybee, head, lateral and anterior views.*

Observe the **haemocoel,** the tubular **heart** with **ostia,** and the various parts of the digestive tract. The digestive tract of insects consists of three main divisions: **foregut, midgut,** and **hindgut.** These divisions are often functionally subdivided for specific functions in particular insects.

The **Malpighian tubules** are outgrowths of the digestive tract originating at the junction of the midgut and hindgut. They serve for excretion. Other internal organs include the **tracheal tubules** and the **gonads.** Is your specimen a male or female? Observe the texture and shape of the gonads and also the posterior segments of the abdomen to determine the sex. Along the ventral surface find the **salivary glands** and the **nerve cord.** Note the **segmental ganglia** and the numerous **lateral nerves.** Observe the powerful muscles of the thorax associated with the legs and wings.

The Honeybee: *Apis mellifera*

The grasshopper was studied as an example of a relatively simple or generalized insect. Many insects, however, have become highly specialized and exhibit numerous adaptations for life in particular habitats or under special conditions. The honeybee is an example of such a specialized insect. Bees are social insects, and they live in permanent colonies organized into three **castes.** Each caste plays a distinct role in the colony and exhibits certain distinctive

morphological features. The **queen** lays the eggs; the **drones** (males) serve only to fertilize the eggs; and **workers** (sterile females) carry on all the other functions of the colony, including constructing and protecting the hive, gathering food, caring for the queen, and rearing the young.

■ Obtain a freshly killed worker for study and examine the specimen for the general characteristics of the phylum, class, and order. Compare the general organization of the bee with that of the grasshopper. How does the surface of the exoskeleton differ from that of the grasshopper? Note the **antennae,** the **compound eyes,** and the three **ocelli** on the head. Locate the **chewing** and the **sucking mouthparts.**

Mouthparts

Remove the head and place it on a microscope slide face-up (figure 13.23). Press on the face and spread the mouthparts with your dissecting needles. Identify the two chewing structures, the **mandibles.** The sucking mouthparts consist of the **labium** or tongue (a long, middle process with a spoonlike structure at the tip), a pair of **labial palps** (one on each side of the labium), and a pair of broad, bladelike structures on the outside, the **maxillae.** Determine how these parts fit together to form a sucking tube. Compare these special sucking mouthparts with the corresponding mouthparts in the grasshopper.

Wings

Examine the wings of the bee and compare them with those of the grasshopper. Study the interlock mechanism, which links the two wings on each side together for flight. Note, also, the network of wing veins, which serves to strengthen the wings.

Legs of the Worker Bee

Consult figure 13.24 and identify the segments of each of the thoracic legs. Starting with the segment next to the body, find the **coxa, trochanter, femur, tibia,** and **tarsus** (foot). The latter is composed of several segments; observe the **tarsal claws** on the terminal segments. The median pad between the claws is the **pulvillus,** which secretes a sticky substance that enables the bee to cling to smooth surfaces. Each of the three legs of the worker exhibits some highly specialized structures; these special adaptations are not present on the appendages of the queen and the drones.

Note on the **first leg** the flattened movable spine, or **velum,** at the distal end of the tibia. This spine, with a notch in the proximal end of the first tarsus, forms the **antenna cleaner.** Note on the opposite side of the velum a number of curved bristles, which form the **pollen brush,**

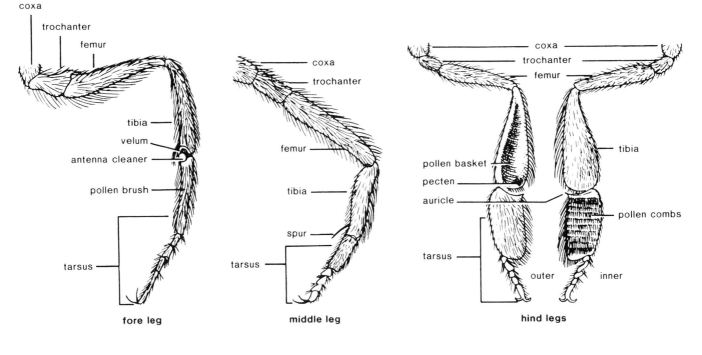

Fig. 13.24 *Honeybee, leg adaptations of worker.*

and at the distal end of the tibia of the **middle leg,** the long **spur,** which is used in picking and transferring wax for comb building.

The third, or **metathoracic leg,** possesses the greatest degree of specialization (figures 13.24 and 13.25). Note the **pollen basket,** a concave depression surrounded by incurving hairs on the external side of the expanded tibia. Note, also, the **pecten,** a row of spines at the end of the tibia, and the **auricle,** a structure opposite the pecten, on the end of the first tarsus. These two adaptations are used to transfer pollen from the pollen combs of the opposite leg to the pollen basket. Find the **pollen combs,** consisting of nine rows of stiff bristles on the inner side of the tarsus. The pollen combs also aid in the removal of wax scales. Observe the wax scales secreted by the wax glands on the ventral side of the abdominal segments.

Abdomen

Count the abdominal segments and compare with the abdomen of the grasshopper. Try to locate the **spiracles.** Observe the sting at the rear of the abdomen. Pull it out slightly with your forceps and note that it consists of a dorsal hollow **sheath** and two **darts.** Observe the **barbs** on the tips of the darts. Which way do they point? What effect would this have on removal of the sting from a wound?

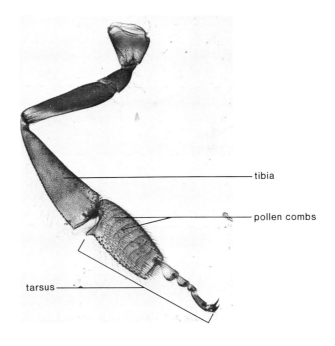

Fig. 13.25 *Grasshopper, hind leg, inner surface. (Courtesy of Carolina Biological Supply Company, Burlington, N.C.)*

Insect Metamorphosis

Many insects undergo marked changes of body form during their postembryonic life so that the immature stages bear little resemblance to the adult stages. These form changes are called **metamorphosis.** Insects differ considerably in the pattern of their postembryonic development, and we can distinguish four general types of postembryonic development.

1. **No metamorphosis.** Some primitive insects (**ametabolous insects**) exhibit no metamorphosis but hatch directly from the egg in a form resembling miniature adults. Examples of this type of development are provided by the silverfish, or bristletails (Order Thysanura), and the springtails (Order Collembola).

2. **Incomplete or gradual metamorphosis.** Young grasshoppers emerge from the egg generally resembling the adult but with a disproportionately large head and without wings. The emergent form is the **nymph,** and transformation to the adult form involves a series of five molts during which the wings are gradually developed and the head-body proportions approach the adult condition. Insects that exhibit incomplete metamorphosis are called **hemimetabolous insects.** In addition to grasshoppers, this type of metamorphosis is exhibited by several other groups of insects including the cicadas (Order Homoptera), walking sticks, mantises, cockroaches (Order Orthoptera), termites (Order Isoptera), and the true bugs (Order Hemiptera).

    Some aquatic hemimetabolous insects have immature forms (naiads) that generally resemble the adults but are adapted for aquatic life. The naiads often have external gills, and they lack wings. They, too, undergo a series of molts and gradually assume the adult form. Dragonflies (Order Odonata), stone flies (Order Plecoptera), and mayflies (Order Ephemeroptera) are examples.

3. **Complete metamorphosis.** Most insects do undergo striking changes in form during their postembryonic life. The characteristic series of stages in complete metamorphosis is: **egg, larva, pupa,** and **adult.** The **egg** hatches into a segmented, wormlike larva. The **larva** feeds, grows, and undergoes several molts, and after the final molt, transforms into a pupa. The **pupa** is a stationary, nonfeeding stage. Although, externally, the pupa appears to be inactive, internally, radical changes are taking place. Most of the larval tissues are resorbed, and the adult organs are newly formed from special groups of embryonic cells called **imaginal discs.** These groups of undifferentiated cells are retained through the larval period and do not resume their development until the pupal stage. When the adult is fully formed within the pupal case, the case is ruptured and the **adult** emerges. Insects demonstrating complete metamorphosis include the beetles (Order Coleoptera), the flies (Order Diptera), the bees and wasps (Order Hymenoptera), and the butterflies and moths (Order Lepidoptera).

■ Study the demonstration materials on insect metamorphosis provided in the laboratory. The grasshopper affords an excellent illustration of incomplete metamorphosis. Observe the eggs and several nymph stages, and compare each of them with the adult. Note especially the changes in the head-body proportions and the gradual development of the wings.

Study complete metamorphosis in a beetle (such as the June beetle, the Mexican bean beetle, or the meal worm *Tenebrio*) and in the silkworm moth (*Bombyx*), or a similar moth. Observe each of the principal stages—the **egg,** the wormlike **larva,** the **pupa,** and the **adult.** On figure 13.26, draw the series of stages for one of the insects studied.

Fig. 13.26 *Student drawing of insect metamorphosis stages.*

Fig. 13.27 Peripatus. *(Courtesy of Carolina Biological Supply Company, Burlington, N.C.)*

Phylum Onychophora

A small and unusual group of animals, the Phylum Onychophora, is frequently cited as a possible evolutionary link between the annelids and the arthropods (figure 13.27). A more recent view advanced by some investigators suggests that the onychophorans may have descended from a primitive segmented ancestral group that also gave rise to both the arthropods and the annelids.

The onychophorans are a small group of about 80 species of small (1.5 to 15 cm long) wormlike animals found in tropical rain forests and other damp or wet habitats in tropical and subtropical areas. Distribution is widespread, but scattered and discontinuous. *Peripatus* is the best-known genus. The caterpillarlike onychophorans exhibit several morphological characteristics similar to those of the annelids and the arthropods. In addition, however, recent investigators have found that they also have several unique features.

The Onychophora are an ancient group that arose in the Paleozoic Era and apparently have changed little since the Cambrian Period. The current belief of most authorities is that long ago the Onychophora evolved separately from the annelids and the arthropods, and that perhaps the three phyla evolved independently from some common ancestral stock.

External Anatomy

Examine a preserved or plastic-embedded specimen of *Peripatus* and observe its general appearance.

The body has a velvety surface that is covered by a thin, chitinous **cuticle.** The body is divided into an anterior **head** and a posterior **trunk. Segmentation** is not apparent from the dorsal side, but note the numerous paired, segmentally arranged legs. The short, stumpy legs bear terminal **claws.**

Study the anterior **head** with one pair of segmented **antennae,** one pair of **simple eyes** (ocelli), and a pair of **oral papillae** just anterior to the **ventral mouth.** Inside the mouth is a pair of sclerotized, hooklike **mandibles** used to hold and to tear prey.

Internal Anatomy

Internally, also, the onychophorans show a strange mixture of annelid and arthropod features (figure 13.28). The body cavity is a **hemocoel,** which contains a simple **tubular digestive tract,** two anterior **slime glands,** a pair of **ventral nerve cords,** the **reproductive organs,** and other internal organs. Respiration is accomplished by a branched **tracheal system** with numerous **spiracles** opening to the outside. The **open circulatory system** consists of a dorsal tubular heart with a pair of openings (ostia) in each segment. Blood flows out the ostia into several **blood sinuses** and back to the heart. A pair of **segmental nephridia** with ciliated funnels open through pores at the base of each leg.

Among the annelid characteristics shown by onychophorans are the muscular body wall, segmental nephridia with ciliated openings, and nonjointed legs. Arthropodlike features include the chitinous cuticle, a reduced coelom, an open circulatory system with a large interior hemocoel, and jaws formed from modified appendages.

Key Terms

Biramous appendage a two-branched appendage characteristic of the crustaceans and certain other invertebrate groups. Consists of a basal portion (protopodite) and two distal branches or rami. The medial branch is the endopodite and the lateral branch is the exopodite.

Book gills respiratory organs consisting of many thin, highly vascularized sheets of tissue. Found in *Limulus.*

Cephalothorax fused head and thorax regions typical of the Crustacea and the Arachnida.

Chelicera pincerlike first appendages characteristic of the Chelicerata. Plural: chelicerae.

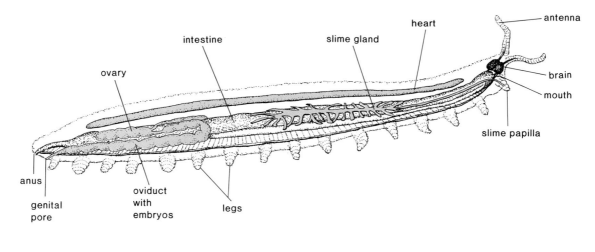

Fig. 13.28 Peripatus, *internal anatomy.*

Cheliped   one of the first pair of walking legs in the crayfish, lobster, and many other large crustacea. Equipped with a terminal pincerlike claw called a chela.

Chitin   a complex structural carbohydrate that forms a chief component of the exoskeleton of many arthropods and other invertebrates.

Compound eye   a type of image-forming photoreceptor with many individual units, the ommatidia. Typical of insects but also found in many crustaceans.

Endopodite   the medial branch (ramus) of the distal portion of a biramous appendage.

Exopodite   the lateral branch (ramus) of the distal portion of a biramous appendage.

Exoskeleton   an external supporting structure or support for the body. Found in arthropods and many other kinds of animals.

Extensor muscle   a muscle that extends or straightens out an appendage or other body part.

Flexor muscle   a muscle that flexes or bends an appendage or other body part toward the body.

Green glands   type of excretory organ found in the crayfish and many other crustacea. Located near the base of the antennae; also called antennal glands.

Hepatopancreas   spongy internal organ of many arthropods, which secretes digestive enzymes, absorbs food, and stores food reserves. Formed as an outgrowth of the midgut.

Larva   an independent, immature stage of an animal morphologically dissimilar to the adult.

Metamorphosis   a change from one form to another in the developmental history of an animal.

Nymph   an immature stage in the life history of insects exhibiting incomplete metamorphosis. Morphologically similar to the adult.

Ocellus   a simple eye found in many invertebrates. Plural: ocelli.

Protopodite   the basal portion of a biramous appendage in arthropods.

Pupa   a dormant stage following the larva stage and preceding the adult stage in insects with complete metamorphosis.

Serial homology   the adaptation of a longitudinal series of originally similar organs to perform different functions, as in the appendages of the crayfish and many other arthropods.

Sexual dimorphism   significant structural differences between the male and female members of a species.

Statocyst   organ of equilibrium found in certain invertebrates, including crustaceans.

Statolith   a sand grain or calcareous granule within a statocyst.

14

Echinodermata

Objectives

After you have completed the laboratory work for this chapter, you should be able to perform the following tasks:

1. List and briefly characterize the five classes of living echinoderms.
2. Identify the principal external features of a starfish. Explain the function of the madreporite, pedicellariae, dermal branchiae, and tube feet.
3. Identify the parts of the digestive tract of a starfish and explain the function of each part.
4. Identify the parts of the water vascular system of a starfish and give the function of each part.
5. Identify the main external features of a sea urchin.
6. Compare the basic organization of a starfish, a sea urchin, and a sea cucumber.

Introduction

The Phylum Echinodermata is a group of spiny-skinned marine animals, which includes the starfish or sea stars, sea urchins, sand dollars, sea cucumbers, brittle stars, and sea lilies (figure 14.1). Some 6,000 species of living echinoderms and more than 20,000 fossil species have been described. Echinoderms are well preserved in the fossil record of ancient times and flourished during several past eras.

Adult echinoderms are easily recognized by their obvious five-part (pentamerous) radial symmetry and calcareous endoskeleton made up of many small plates, which may be separate (as in the sea cucumbers) or fused into a rigid framework or test (as in a sea urchin or sand dollar). In many of the echinoderms, these mesodermally derived skeletal plates bear protruding spines. Echinoderms have

a large coelom lined with a ciliated peritoneum, a portion of which forms into the unique water vascular system during embryonic development. The water vascular system is of great importance to the echinoderms and plays a vital role in their locomotion, attachment, respiration, food handling, and sensory perception.

The echinoderms are of special interest because of their close evolutionary relationship to the Phylum Chordata. Strong embryological evidence suggests a close link between the echinoderms and the chordates. Their mesodermal endoskeleton, the formation of the coelom from outpockets of the archenteron (primitive gut) in the embryo, the formation of the anus from the embryonic blastopore, and the radial indeterminate cleavage of the early embryo are all chordatelike characteristics.

Classification

Some disagreement exists among specialists on the classification of the Phylum Echinodermata, but most recent workers have adopted the following classification of living echinoderms, with three subphyla and five classes represented among the present North American fauna.

Subphylum Echinozoa

Nonsessile animals with rounded or disc-shaped body without arms.

Class Echinoidea

Sea urchins, heart urchins, and sand dollars. Bottom-dwelling (benthic) echinoderms with a rigid test of fused skeletal plates. Body covered with movable spines. Five rows of tube feet (bearing suckers) around the test. Examples: *Arbacia, Lytechinus, Strongylocentrotus* (all sea urchins); *Echinarachnius, Mellita,* and *Dendraster* (sand dollars).

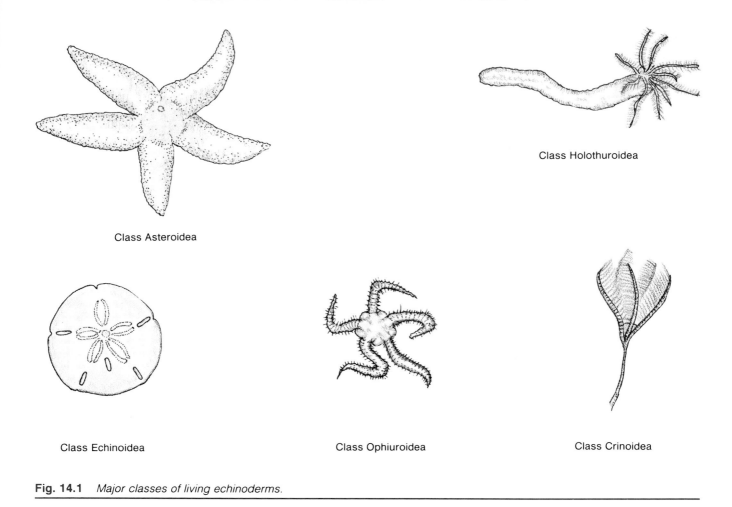

Class Asteroidea

Class Holothuroidea

Class Echinoidea

Class Ophiuroidea

Class Crinoidea

Fig. 14.1 *Major classes of living echinoderms.*

Class Holothuroidea

Sea cucumbers. Soft-bodied animals having a flexible body wall with many tiny embedded calcareous ossicles; no spines or arms. Body elongated in the oral-aboral axis to a wormlike form; a circle of tentacles surround the mouth. Usually with five or more series of tube feet arranged in longitudinal rows. Examples: *Thyrone, Cucumaria, Stichopus.*

Subphylum Crinozoa

Spherical or cup-shaped body with branching arms; attached to the substrate during all or part of life by a stem; oral surface directed upward.

Class Crinoidea

Sea lilies, basket stars. Flowerlike echinoderms with a central calyx and five (or multiples of five) branching arms. Ciliated ambulacral groove along the oral surface of the arms serves for food gathering. Some species attach to the sea bottom by a stalk, others free-swimming. Examples: *Cerocrinus* (crinoid), *Hathrometra* (=*Antedon*) (feather star).

Subphylum Asterozoa

Nonsessile, star-shaped echinoderms, with radial arms and oral surface directed downward.

Class Asteroidea

Starfish or sea stars. Animals with a star-shaped body, typically with five arms and a flexible skeleton with many calcareous plates. The arms are continuous with and not clearly distinct from the central disc. With ambulacral grooves on the oral side of each arm; tube feet bearing suckers. Examples: *Asterias, Pisaster, Astropecten.*

Class Ophiuroidea

Brittle stars. Star-shaped echinoderms with arms distinct from the central disc; no ambulacral grooves along the arms; tube feet absent or reduced to sensory organs. Examples: *Amphipholis, Ophiothrix* and *Ophioderma* (brittle stars); and *Gorgonocephalus* (Pacific basket star).

Materials List

Preserved specimens
  Asterias
  Arbacia
  Cucumaria or *Thyone*
  Aristotle's Lantern (Demonstration)
  Starfish, dried to show ossicles (Demonstration)
  Representative echinoderms (Demonstration)
Prepared microscope slides
  Starfish, pedicellaria (Demonstration)
  Starfish, dermal branchiae (Demonstration)
  Starfish, bipinnaria larva (Demonstration)
  Starfish, metamorphosis (Demonstration)

The Common Starfish: *Asterias*

Class Asteroidea

Asterias forbesi is a common starfish (see figure 14.2) found along the Atlantic coast of the United States. Obtain a living or preserved specimen of *Asterias* and observe its radial symmetry. The radial symmetry of adult echinoderms appears to be a secondary condition, since echinoderm larvae are bilaterally symmetrical. Observe a demonstration slide of the bipinnaria larva of the starfish. In what ways does the radial symmetry of the echinoderms differ from that of the coelenterates?

External Anatomy

Specimens preserved for laboratory study often are injected with a colored material to facilitate the study of the water vascular system, including the external tube feet.

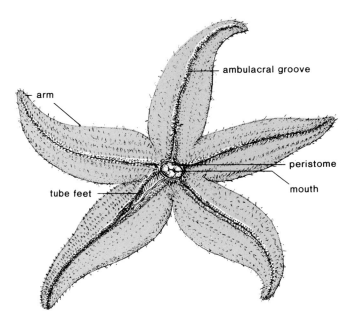

Fig. 14.2 *Starfish, oral view.*

Keep your specimen wet by adding a little water to the dissecting pan. Note the **central disc** on the upper, or **aboral,** side; the five rays or **arms;** and the **madreporite,** a light-colored circular area near the edge of the disc at the junction of two rays. The madreporite is a calcareous disc that serves as the entrance to the water vascular system. The two arms adjacent to the madreporite comprise the **bivium;** the other three arms make up the **trivium.** The arm opposite the madreporite is referred to as the **anterior** arm.

Note the many spines scattered over the surface of the arms and the central disc. Among the spines there are also many small pincerlike structures located near the bases of the spines (see demonstration slide). These structures are the **pedicellariae.**

■ Carefully scrape a small area of the arm with your scalpel, add the scrapings to a drop of water, and observe the material under the low power of your microscope. Can you identify any pedicellariae among the scrapings? From their structure, what do you think the function of the pedicellariae might be?

Fig. 14.3 *Student drawing of central disc and one arm of starfish, oral view.*

Also among the spines are numerous **papulae** (dermal branchiae), soft, hollow projections of the body wall used for respiration (see demonstration slide). Observe the skeletal calcareous plates, or **ossicles,** on a dried starfish specimen.

Note on the oral side of the starfish: the **mouth** guarded by specialized **oral spines** and surrounded by a soft membrane, the **peristome;** the five **ambulacral grooves,** extending from the mouth and along the middle of each arm; and the numerous **tube feet** in the grooves, extending from the water vascular system. There is also a pigmented **eyespot** and a **sensory tentacle** at the tip of each ray, but these latter structures are usually difficult to observe in preserved specimens.

■ Draw an enlarged view of the central disc and one arm of a starfish (in figure 14.3). Show an oral view and include the following structures in your drawing: **mouth, peristome, oral spines, ambulacral groove,** and **tube feet.** Label each structure.

Internal Anatomy

■ Carefully examine figures 14.4 and 14.5 for orientation before starting your dissection. Cut off about one-half inch from the tip of the anterior arm. Then carefully cut along each side of the anterior arm to its junction with the central disc along each side and across the top of the arm at the margin of the disc. Observe the hard, calcareous plates, or ossicles, of the skeleton as you cut. Remove the portion of the body wall covering the aboral surface of the arm and examine the large **coelom** within the arm, which contains the internal organs. The coelom is lined with a ciliated **peritoneum,** as mentioned previously, and is normally filled with coelomic fluid. This fluid carries oxygen and dissolved nutrients to various parts of the body. Next, remove the portion of the body wall covering the aboral side of the central disc, but leave the madreporite in place by carefully cutting around it.

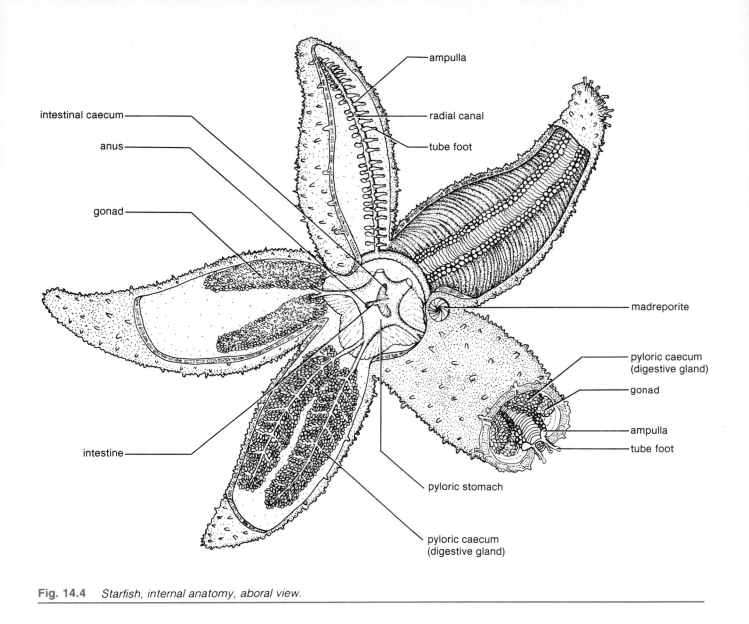

intestinal caecum

anus

gonad

intestine

ampulla

radial canal

tube foot

madreporite

pyloric caecum
(digestive gland)

gonad

ampulla

tube foot

pyloric stomach

pyloric caecum
(digestive gland)

Fig. 14.4 *Starfish, internal anatomy, aboral view.*

Digestive System

The starfish has a complete digestive system. The mouth of the starfish is on the oral surface and opens into a short esophagus. The esophagus leads to a large, two-chambered stomach that fills most of the central disc. The large, thin-walled cardiac chamber lies below (oral to) and a smaller, thick-walled chamber lies above (aboral to) the cardiac chamber.

Emptying into the pyloric stomach through five pyloric ducts are five pairs of large, branched pyloric caecae (digestive glands). The pyloric caecae fill most of the in-

terior space in the arms. Also leading from the pyloric stomach is a small, short intestine that empties through the aboral anus.

The cardiac chamber can be everted through the mouth to envelop its prey. Most starfish are carnivores and feed on many kinds of invertebrates, including clams, oysters, gastropods, crustaceans, polychaetes, other echinoderms, and small fish. Prey is often engulfed whole, including shell or exoskeleton, and the undigestible parts are later regurgitated. Typically, food is digested in the cardiac chamber of the stomach. Digestive enzymes are secreted principally by the pyloric caecae, which also play a major role in absorption and storage of food materials.

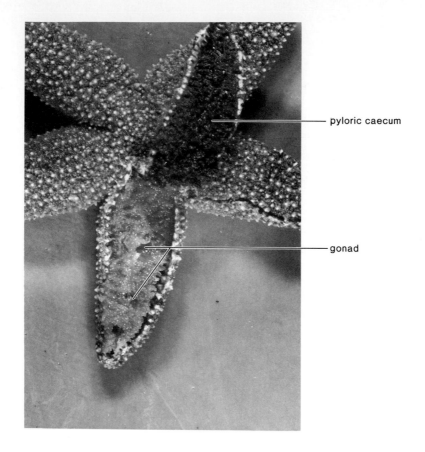

pyloric caecum

gonad

Fig. 14.5 *Starfish, partly dissected, aboral view. (Photograph by Ken Taylor.)*

Water Vascular System

The water vascular system (figure 14.6) is a unique feature of the echinoderm. It develops embryologically as a specialized part of the coelom and is lined with a **ciliated epithelium.** The water vascular system has several important functions, including locomotion, feeding, respiration, excretion, and sensory perception. The canals of this system are filled with a water vascular fluid that has a chemical composition resembling seawater plus some soluble proteins and an elevated level of potassium ions. Suspended in the water vascular fluid are numerous **coelomocytes,** amoeboid cells that serve in defense, excretion, and perhaps in several other roles as well.

Starfish have a well-developed water vascular system. Among its principal parts is the external **madreporite,** seen previously on the aboral surface, which marks the entrance to a short **stone canal** (see figure 14.6). The madreporite is covered by the thin, ciliated epidermis of the body surface, and is believed to function as a pressure regulator for the water vascular system rather than as a

simple filter, as has sometimes been assumed. The stone canal gets its name from the calcareous deposits surrounding it. The stone canal leads to the **circular canal,** which encircles the mouth. Five **radial canals** lead from the circular canal along the top of the **ambulacral groove** leading into each arm. Many short **lateral canals** connect the radial canals with each pair of **tube feet.** Each tube foot consists of a bulblike **ampulla** attached to each suckerlike tube foot.

Within each lateral canal is a valve that closes to allow local pressure changes in the individual tube feet. Contraction of the circular muscles in the wall of the ampulla increases the internal pressure and extends the foot. Contraction of longitudinal muscles in the wall of the tube foot withdraws the foot. These pressure changes within the water vascular system plus nervous control from the central nervous system allow coordinated movements of the tube feet and, thus, coordinated locomotion of a starfish.

Tiedeman's bodies

Polian vesicles
madreporite
stone canal

radial canal

circular canal

ampulla

tube foot

sucker

Fig. 14.6 *Water vascular system of a starfish.*

Nervous System

The starfish has a very simple nervous system. A circular **nerve ring** surrounds the mouth, and a **radial nerve** extends from the nerve ring into each arm. A simple light-sensitive **eyespot** at the tip of each arm is the only differentiated sense organ, but sensory cells are scattered throughout the epidermis.

Respiration and Excretion

Respiration is carried on through the thin walls of the **papulae** (dermal branchiae) extending from the body surface and also through the walls of the tube feet. Excretion of nitrogenous wastes is accomplished mainly by diffusion of ammonia and urea from the respiratory surfaces and the body wall. Coelomocytes engulf insoluble particulate matter and carry these wastes to the surface where they are discharged. Echinoderms have no differentiated excretory organs.

Reproductive System

The reproductive system consists of a pair of **gonads** located near the base of each arm (see figure 14.4). The sexes are separate in sea stars, but are difficult to distinguish except by microscopic examination of the gonad contents.

■ Take a small amount of tissue from one of the gonads and macerate it in a drop of water on a clean microscopic slide. Can you identify any large oocytes (eggs) or any flagellated sperm?

From each gonad a duct leads to an external pore on the aboral surface. Eggs and sperm are released by mature sea stars into the seawater, where fertilization occurs, and the zygotes develop into free-swimming **bipinnaria larvae.**

Fig. 14.7 *Sea urchin,* Arbacia. *(Courtesy of Carolina Biological Supply Company, Burlington, N.C.)*

A Sea Urchin

Class Echinoidea

Sea urchins are bottom-dwelling (benthic) echinoderms with a **rigid endoskeleton** or **test,** consisting of many fused plates or ossicles. They lack arms and typically are globular or ovoid in shape, and their bodies are covered with movable **spines** (figure 14.7). Among the spines are many tube feet and numerous **pedicellariae,** which serve to keep the test clean of debris and fouling organisms.

◼ Study a living or preserved specimen of *Arbacia* or a similar sea urchin. Identify the rigid **test, spines, dermal branchiae,** and **pedicellariae.** Locate also the five series of **tube feet** along five symmetrically placed **ambulacral regions** around the globular body. In a living specimen, the tube feet can extend well beyond the spines.

   On the oral side of the urchin (flattened surface), find the **mouth** surrounded by five **calcareous teeth** and a circular oral membrane, the **peristome.** Five pairs of peristomial gills are located around the peristome. The teeth are connected internally to a complex chewing apparatus, called **Aristotle's Lantern,** which anchors the teeth and enables them to scrape algae from submerged rocks. Urchins are principally herbivorous, and many species feed on algae from the sea floor; they are sometimes called the grazers of the sea. They live mostly along rocky shores. Their relatives, the sand dollars and heart urchins, are adapted for life on soft sea bottoms in which they burrow in search of food and shelter.

Some Common Echinoids

Arbacia punctulata—purple sea urchin found along the Atlantic coast from Cape Cod to Florida.

Strongylocentrotus drobachiensis—green sea urchin found in colder waters north of Cape Cod along the western Atlantic, in northern Europe, and from Washington north to Alaska on our Pacific coast.

Strongylocentrotus purpuratus—the Pacific purple sea urchin found on surf-swept rocks from California to Alaska.

Echinarachnius parma—Northern Atlantic sand dollar found on sandy areas on the sea bottom from Labrador to New Jersey, and also in the Northern Pacific from British Columbia across to Japan.

Mellita quinquiesperforata—Keyhole sand dollar abundant from Cape Hatteras to the Caribbean.

A Sea Cucumber

Class Holothuroidea

Sea cucumbers are soft-bodied echinoderms with many tiny plates or ossicles embedded in their **leathery body wall.** They are principally filter feeders, collecting plankton with their tentacles, or deposit feeders, collecting food materials mixed with the bottom sediments.

◼ Study a living or preserved *Cucumaria* (figure 14.8) or *Thyone* and identify the **oral tentacles,** the external **tube feet,** and the **leathery body wall.**

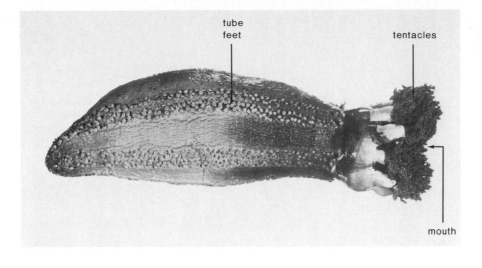

tube feet

tentacles

mouth

Fig. 14.8 *Sea cucumber,* Cucumaria. *(Courtesy of Carolina Biological Supply Company, Burlington, N.C.)*

Some Common Holothuroidea

Thyone briareus—burrowing sea cucumber found in shallow waters on soft, sandy, or muddy sea bottoms from Cape Cod to the Gulf of Mexico. Tube feet scattered over body surface.

Cucumaria frondosa—deeper water, North Atlantic species in which the tube feet are arranged along five distinct ambulacral areas.

Leptosynapta tenuis—long, wormlike sea cucumber, modified for burrowing in soft bottom sediments. Lacks regular tube feet.

Demonstrations

1. Pedicellariae (microscope slide).
2. Dermal branchiae (microscope slide).
3. Bipinnaria larva (microscope slide).
4. Starfish metamorphosis (microscope slide or color transparency series).
5. Preserved or dried Aristotle's Lantern from sea urchin.
6. Dried starfish to show ossicles.
7. Preserved specimens representing other types of echinoderms, such as crinoids, basket stars, and brittle stars.

Key Terms

Papulae (dermal branchae) hollow projections of the body wall that function in respiration in starfish and other asteroids.

Pedicellariae small, pincerlike structures found on the surface of many starfish (Class Asteroidea), sea urchins (Class Echinoidea), and brittle stars (Class Ophiuroidea). Serve to keep the body surface free from debris and fouling organisms.

Radial symmetry type of body organization in which all parts of the body are arranged symmetrically around a central axis.

Test a rigid endoskeleton composed of many fused calcareous plates as exhibited by the sea urchins.

Water vascular system a unique organ system found only in the Phylum Echinodermata. Functions in feeding, attachment, locomotion, respiration, and sensory perception. Formed embryologically as a specialized portion of the coelom.

15

Chordata

Objectives

After completing the laboratory work in this chapter, you should be able to perform the following tasks:

1. Describe the life cycle of a tunicate (Subphylum Urochordata) and identify the principal features of a tunicate tadpole larva. Explain the morphological and evolutionary significance of the larva.

2. Identify the main organs in an adult tunicate and discuss the adaptations of the adult for its sessile mode of life.

3. Explain the feeding mechanism of a tunicate and identify the principal structures involved.

4. Identify the main organs in a lancelet and discuss the significance of its basic chordate features.

5. Compare the organization of a tunicate and a lancelet and explain their main similarities and differences.

Introduction

The Phylum Chordata includes a remarkable range of forms, varying from relatively simple marine animals to highly specialized birds and mammals. Despite the wide range of diversity in the phylum, however, all members exhibit three fundamental chordate characteristics during some stage in their life histories. These distinctive chordate features are (1) a dorsal hollow nerve cord; (2) a notochord, a dorsal, elastic supporting rod; and (3) paired pharyngeal gill slits. Some zoologists add a fourth characteristic, a post-anal tail. Nonetheless, the chordates comprise a large, diverse, and distinct phylum of animals that have successfully populated the land, the waters, and the air. This large and important phylum is commonly divided into two groups and three subphyla.

Classification

Group Acrania (Protochordata)

Chordates without a cranium or braincase.

Subphylum Urochordata (Tunicates or Sea Squirts)

Animals with a well-developed notochord and dorsal nerve cord in the free-swimming larva; specialized adults, sessile or planktonic, and lacking a notochord and dorsal nerve cord. Examples: *Molgula* (sea grape), *Styela*, *Amaroucium* (sea pork).

Subphylum Cephalochordata (Lancelets)

Elongate, fishlike chordates with a persistent notochord and dorsal nerve cord. Example: *Branchiostoma* (*Amphioxus*).

Group Craniata

Chordates with a cranium enclosing the brain and sense organs of the head; exhibit substantial cephalization.

Subphylum Vertebrata (Vertebrates)

Chordates with a backbone, skull, brain, and kidneys.

Class Agnatha (Lampreys). Fishlike vertebrates without jaws or appendages. Example: *Petromyzon.*

Class Elasmobranchiomorphii (Chondrichthyes) (Sharks and Rays). Cartilaginous fishes, usually with external placoid scales. Examples: *Squalus* (shark), *Raja* (skate), *Chimaera* (ratfish).

Class Osteichthyes (Bony Fishes). Fishes with skeleton primarily of bone, usually with swim bladder or lungs. Examples: *Amia* (bowfin), *Perca* (perch), *Lepistoseus* (gar), *Micropterus* (bass).

Class Amphibia (Salamanders, Frogs, and Toads). Four-limbed vertebrates (tetrapods) with a soft, moist skin; three-chambered heart; eggs enclosed in gelatinous covering; fertilization and development usually restricted to fresh water. Adults usually aquatic or semiaquatic. Examples: *Rana* (frog), *Bufo* (toad), *Ambystoma* (salamander).

Class Reptilia (Lizards, Snakes, Turtles, and Alligators). Four-limbed vertebrates with a dry, cornified skin; eggs enclosed in a protective shell resistant to drying (amniote egg); four-chambered heart. Examples: *Anolis* (lizard), *Chrysemys* (turtle), *Crotalus* (rattlesnake), *Alligator.*

Class Aves (Birds). Winged vertebrates with feathers and constant high body temperature. Examples: *Sturnus* (starling), *Cyanocitta* (blue jay), *Corvus* (crow).

Class Mammalia (Mammals). Warm-blooded vertebrates; body covered with hair; young nourished by milk produced by female. Examples: *Homo* (human), *Sus* (pig), *Equus* (horse), *Canis* (dog).

In this chapter we will study representatives of the first two subphyla, the urochordates and the cephalochordates.

Materials List

Preserved specimens
  Molgula
  Branchiostoma
  Representative tunicates
Prepared microscope slides
  Amaroucium, tadpole larva
  Branchiostoma, whole mount
  Branchiostoma, cross section

Subphylum Urochordata, Tunicates or Sea Squirts

Urochordates are interesting, important, and unusual animals. They are interesting and important because they represent the simplest living chordates. About 1,250 species are currently recognized. They are unusual because adult urochordates became adapted to a special mode of life as sedentary, filter-feeding, marine animals, and, in the process, have lost two of the three basic chordate features. Larval urochordates, however, have retained all three chordate features and clearly establish the urochordates as legitimate ancestors of the higher chordates, including fishes, birds, mammals, and humans.

The Tunicate Larva

The larval stages of tunicates are of special importance because they clearly exhibit the three fundamental chordate characteristics: tubular nerve cord, notochord, and pharyngeal gill slits. The first two of these features are lost in the adult tunicate, presumably because of the specialization of the adult due to its sessile mode of life.

Tunicate larvae are often called **tadpole larvae** because of their superficial resemblance to the tadpole larvae of amphibians.

■ Study a prepared microscope slide with a whole mount of the larva of *Amaroucium,* or a similar tunicate larva, and observe its characteristic form (see figure 15.1).

Note the **muscular tail** and thickened body. At the anterior end of the body find the adhesive papillae by means of which the larva attaches to the substrate prior to metamorphosing into a sessile adult. Near the dorsal surface is a **sensory vesicle** with two conspicuously pigmented sense organs—a light-sensory **ocellus**, or eye-spot, and a **statolith** that serves as a balancing organ.

Anterior to the darkly pigmented sensory vesicle, locate the **incurrent siphon** through which water is pumped into the pharynx. The **excurrent siphon** is found slightly posterior to the sensory vesicle, near the attachment of the tail. Locate also the **gill slits** in the wall of the pharynx.

The **notochord**, the **dorsal nerve cord,** and conspicuous **muscle bands** are present in the tail. At the time of metamorphosis, the larva settles to the bottom, attaches by means of a secretion of glands in the adhesive papillae,

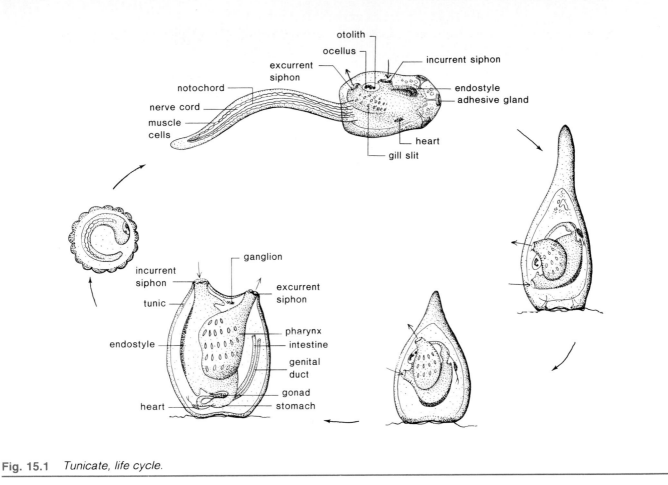

Fig. 15.1   *Tunicate, life cycle.*

and the tail is partly resorbed. The notochord, nerve cord, and muscles are broken down, and several new organs appear in the body of the tunicate to complete the transformation into the adult. The life cycle of a tunicate is illustrated in figure 15.1.

The Adult Tunicate

Adult tunicates may be solitary or colonial, sessile or planktonic. The most familiar ones, however, are the sessile forms found attached to rocks, jetties, and pilings in shallow seas (figure 15.2).

The typical structure of an adult tunicate is well illustrated by *Molgula,* the sea grape, found in many places along the Atlantic coast.

■ Study the demonstration of a partially dissected specimen of *Molgula* and note the tough, fibrous outer covering called the tunic (see figure 15.3). This outer covering is unusual because it is composed partly of cellulose, a carbohydrate normally produced only by plants.

Projecting from the globular body are two siphons: the incurrent siphon is anterior, and its opening is divided into six lobes. The more posterior excurrent siphon has a square opening. Locate internally the large pharynx with numerous gill slits. Ventral to the pharynx are the short esophagus, the bulbous stomach, and the beginning of the intestine. The intestine continues dorsally and ends at the anus just below the excurrent siphon.

Molgula, like most other tunicates, is a specialized filter-feeder. Water is drawn in through the incurrent siphon by the action of cilia lining the internal chambers. The water is drawn into the large pharynx, and minute organisms suspended in the water (plankton) are trapped in the mucus on the pharyngeal walls as the water passes out through the many gill slits.

Special ciliary currents collect the food within the pharynx and pass it into the esophagus and on to the stomach. Water leaving the pharynx passes into the atrium, a large cavity surrounding the pharynx, and from the atrium the water flows out through the excurrent siphon.

Fig. 15.2 Cynthia. *A colonial tunicate. (Photograph courtesy of Carolina Biological Supply Company, Burlington, N.C.)*

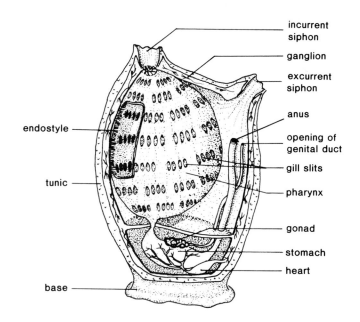

incurrent siphon

ganglion

excurrent siphon

anus

opening of genital duct

gill slits

pharynx

gonad

stomach

heart

endostyle

tunic

base

Fig. 15.3 *Tunicate, adult.*

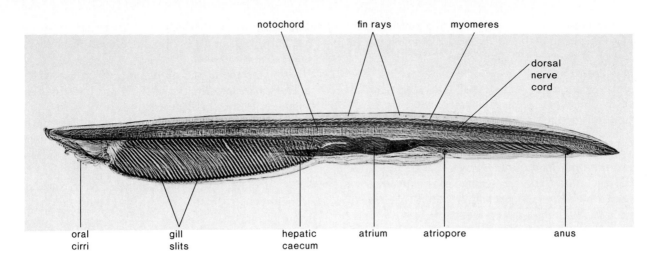

notochord   fin rays   myomeres

dorsal
nerve
cord

oral
cirri   gill
slits   hepatic
caecum   atrium   atriopore   anus

Fig. 15.4 *Lancelet, whole mount. (Courtesy of Carolina Biological Supply Company, Burlington, N.C.)*

The **endostyle,** a ciliated groove located on one side of the large pharynx, is of special importance. The endostyle secretes a stream of mucus that is carried along by the motion of the cilia to form a sheet that traps small food particles from the incoming seawater. The mucus sheet with its captured food particles is then passed to the esophagus and into the stomach and through the digestive system. The endostyle has been found to be the forerunner of the vertebrate thyroid gland.

The circulatory system in *Molgula,* as in most tunicates, consists of a ventral heart located near the stomach. Attached to the heart are two large vessels that carry blood to other organs. The tunicate heart is unique in that it provides two-way propulsion. The heart first pumps blood in one direction, then reverses and pumps blood in the opposite direction.

In the adult tunicate, the dorsal hollow nerve cord and the notochord are absent. Their loss is generally believed to represent a specialization of the adult tunicate for its peculiar sessile mode of life.

A Lancelet: *Branchiostoma*

Subphylum Cephalochordata

The Subphylum Cephalochordata consists of a few species of small fishlike animals called lancelets (see figure 15.4), which live in shallow seas in many parts of the world. In certain areas, the lancelets are sufficiently abundant that they are used as human food. The common American lancelets are traditionally called *Amphioxus* (formerly a generic name), although most specialists now use the generic name *Branchiostoma.*

External Anatomy

◼ Study first a preserved specimen and observe its general form and external features. Then obtain a prepared microscope slide with a stained whole mount of a lancelet to study its internal anatomy.

Mature lancelets are usually between two and three inches in length, but smaller, immature forms are normally used in making microscopic whole mounts.

In the preserved specimen, note the slender, elongate shape of the animal, the absence of a distinct head, and the lack of paired fins or limbs.

■ Handle the specimen with care and do not dissect it. Return it intact to the proper container when you have completed your study.

Refer to figure 15.5 and identify the anterior **rostrum** and the **oral hood** bordered by a fringe of ciliated **oral cirri** enclosing the large **vestibule.** The **mouth** is an opening in a membrane, the **velum,** located at the rear of the vestibule. Surrounding the mouth are several **velar tentacles.** How many? Cilia arranged in bands along the walls of the vestibule form the "wheel organ," which generates water currents and carries seawater containing suspended food organisms into the mouth.

Internal Anatomy

Behind the mouth is a large pharynx with many diagonal **gill slits** on each side. Between the gill slits are **gill bars,** each supported by a thin cartilaginous rod. The pharynx plays an important role both in feeding and in respiration, but its primary function is in feeding. The respiratory function is clearly secondary. Posterior to the pharynx is a straight intestine, which ends at the subterminal anus. A slender pocket, the **midgut caecum** (believed by some workers to be homologous with the vertebrate liver), opens on the ventral side of the intestine near the junction of the pharynx and intestine and extends forward.

Locate the **dorsal, caudal** (tail), and **ventral fins.** Note the short **fin rays** composed of connective tissue within the fins. Observe the **atriopore,** a midventral opening located anterior to the ventral fin. Water taken into the pharynx passes out through the gill slits into the **atrium** and out of the atrium via the **atriopore** (see figure 15.5).

Lancelets, like the tunicates studied earlier, are filter-feeders. Seawater containing planktonic organisms is drawn by ciliary currents through the mouth into the pharynx where food particles are trapped in mucous secretions. The mucus, containing trapped food particles, is swept posteriorly to the intestine, and the water passes out of the pharynx through the lateral gill slits. Respiratory exchange also occurs as the water passes through the gill slits and past the gill bars, which contain blood vessels.

Note the conspicuous V-shaped structures along each side of the body; these are the muscle segments, or **myomeres.** Contraction of these muscles produces a lateral

bending of the body, which aids the lancelet in swimming and burrowing in the bottom sediments where it commonly dwells.

Locate the dorsal **notochord,** which extends longitudinally just dorsal to the pharynx and intestine. The notochord is a thin, cartilaginous rod surrounded by a sheet of connective tissue. The contraction of the myomeres against the rigidity of the notochord produces the lateral swimming movements of the body that propel the lancelet forward. Find the **dorsal nerve cord** just ventral to the dorsal fin and dorsal to the notochord. At its anterior end is a slight enlargement, the **cerebral vesicle,** a very primitive sort of "brain."

The circulatory system of *Branchiostoma* consists of a network of elastic vessels similar to that found in the higher chordates, but it lacks a distinct heart. It is difficult to study the details of the circulatory system except by the dissection of specially prepared specimens, but portions of the circulatory system can be observed in the microscopic cross sections to be studied later. Figure 15.7 shows the principal blood vessels and the general pattern of circulation in a lancelet.

Blood from the digestive tract is collected by the **subintestinal vein,** which leads to the **hepatic portal vein,** which, in turn, carries the blood to the midgut caecum. The **hepatic vein** leaves the liver and leads to the **ventral aorta** below the pharynx. Numerous **afferent branchial arteries** (each with a contractile bulb at its base) branch from the ventral aorta and carry the blood upward to the gill bars where it is oxygenated. Pulsations of the ventral aorta and of the enlargements at the bases of the afferent branchial arteries appear to aid in pumping the blood through the system. From the gills, the blood is transported by the **efferent branchial arteries** and is collected in the **paired dorsal aortas** (right and left) above the gills. The two dorsal aortas (see figure 15.6) join posteriorly to form a single **median dorsal aorta** just behind the pharynx. This latter vessel carries oxygenated blood posteriorly to the body tissues and the intestine to complete the circuit.

Cross Sections

Prepared microscopic cross sections can help greatly to supplement your observations on the preserved specimens and whole mounts and to improve your understanding of the anatomy of amphioxus.

■ Study several cross sections from different regions of the body and attempt to identify as many of the internal and external structures as possible.

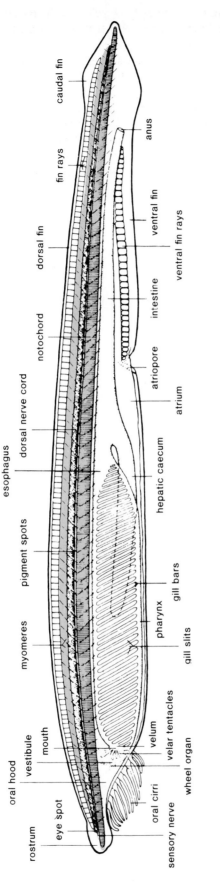

Fig. 15.5 *Lancelet, internal organs.*

Labels (Fig. 15.5): caudal fin, fin rays, anus, ventral fin, ventral fin rays, dorsal fin, intestine, notochord, atriopore, atrium, dorsal nerve cord, esophagus, hepatic caecum, pigment spots, gill bars, myomeres, pharynx, gill slits, velum, velar tentacles, mouth, wheel organ, vestibule, oral hood, oral cirri, sensory nerve, eye spot, rostrum

Fig. 15.6 *Lancelet, circulatory system. Small arrows in blood vessels show direction of blood flow.*

Labels (Fig. 15.6): dorsal fin, notochord, caudal fin, anus, intestine, subintestinal vein, median dorsal aorta, intestinal capillaries, atriopore, segmental arteries, atrium, paired dorsal aorta, hepatic vein, hepatic portal vein, efferent branchial arteries, liver capillaries, dorsal nerve cord, ventral aorta, gill slits, gill bars, afferent branchial arteries, velar tentacles, mouth, eye spot, oral cirri, sensory lobes

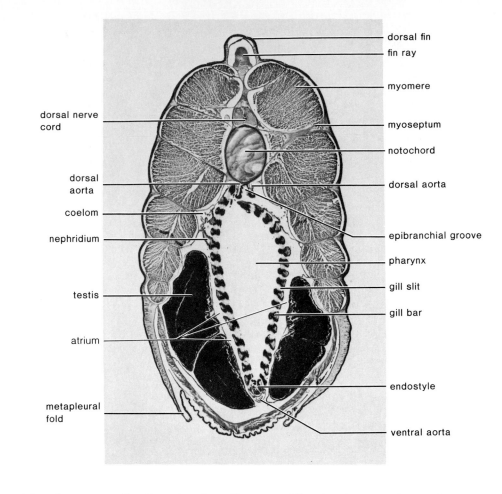

Labels on the figure:
- dorsal fin
- fin ray
- myomere
- myoseptum
- notochord
- dorsal aorta
- epibranchial groove
- pharynx
- gill slit
- gill bar
- endostyle
- ventral aorta
- dorsal nerve cord
- dorsal aorta
- coelom
- nephridium
- testis
- atrium
- metapleural fold

Fig. 15.7 *Lancelet, male, cross section through atrium. (Courtesy of Carolina Biological Supply Company, Burlington, N.C.)*

The following description should aid you in identifying various structures in the cross section. It is based upon a cross section through the pharyngeal region as illustrated in figure 15.7.

Observe the two-layered **skin,** with an outer **epidermis** consisting of a single layer of columnar epithelium, and an underlying **dermis,** a thin layer of gelatinous connective tissue. Find the **dorsal fin** and **fin ray,** the **myomeres** bounded by the myosepta of connective tissue, the **notochord,** and the **dorsal nerve cord** beneath it. Locate the **central canal** within the dorsal nerve cord. Observe also the nerve cells and fibers of the dorsal nerve cord and the **spinal nerves** (not present in every section because of differences in the plane of sectioning).

Observe the laterally compressed pharynx and the numerous **gill slits** between the **gill bars.** The ventral ciliated groove is the **endostyle,** or hypobranchial groove, and the dorsal ciliated groove is the **epipharyngeal groove.** Both play an important role in the trapping and transporting of food particles. There is good evidence that the endostyle is homologous with the vertebrate thyroid gland.

The chamber around the pharynx is the **atrium.** What tissue layer forms the lining of the atrium? What tissue layer lines the coelom? Note the **gonads** extending into the atrium and the section of the midgut caecum also extending into it. The paired cavities above and lateral to the pharynx are portions of the **coelom,** as is the small cavity found below the endostyle. Locate the **blood vessels** in your section, including the **paired dorsal aortae** above the pharynx and the **single median ventral aorta** below it. Find, also, the **hepatic vein** or veins closely associated with the midgut caecum. The **nephridia** are small ciliated ducts

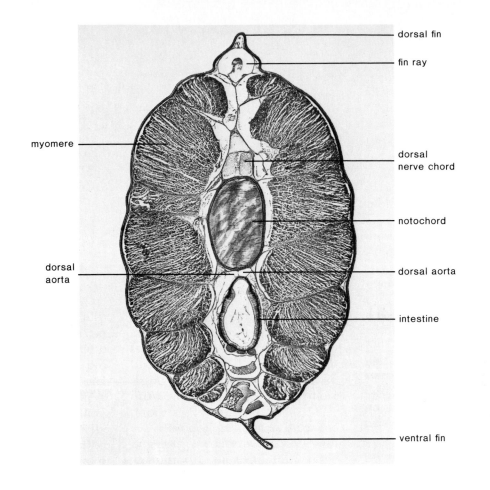

Fig. 15.8   *Lancelet, cross section through intestinal region. (Courtesy of Carolina Biological Supply Company, Burlington, N.C.)*

that connect the dorsal portions of the coelom with the atrium. The two **metapleural folds** at the two sides on the ventral surface of the body should be apparent.

Compare the structures observed in the cross section through the pharyngeal region with a cross section through a more posterior region, as illustrated in figure 15.8.

Demonstrations
1. Preserved lancelets.
2. Examples of solitary and colonial tunicates.

Key Terms

Endostyle   a ciliated groove located on the ventral surface of the pharynx in tunicates and lancelets. Functions in the capture of food particles. Possibly homologous with the thyroid gland of vertebrates.

Gill slits   paired openings in the lateral walls of the pharynx in urochordates and cephalochordates. Important in food capture and secondarily in respiration.

Myomeres   segmented muscle blocks arranged longitudinally along the dorsal portion of the lancelet body. Contraction of the myomeres, combined with the relatively stiff notochord, causes the lancelet body to flex, and produces effective swimming and burrowing movements.

Notochord   a stiff supporting cartilaginous rod of mesodermal origin found in the tunicate tadpole and in adult cephalochordates. Absent in adult tunicates.

Tadpole larva   characteristic larval form of the tunicates with a notochord, dorsal tubular nerve cord, and paired gill slits. Superficially resembles an amphibian tadpole larva.

16

Frog Anatomy

Objectives

After completing the laboratory exercise for this chapter, you should be able to complete the following tasks:

1. Identify the main structures within the oral cavity of the frog and explain the function of each.
2. Describe the external features (secondary sexual characteristics) that distinguish mature male and female frogs.
3. Describe the major divisions of the frog skeleton and explain the chief functions of each division.
4. Describe the principal parts of the digestive system of the frog and explain their function.
5. Describe the structure of the frog heart and identify its major parts.
6. Describe the basic pattern of blood circulation in the frog and identify the principal blood vessels involved.
7. Identify the important parts of the urogenital system of both male and female frogs, and explain their functions.
8. Identify the five main regions of the frog brain and the components of each. Briefly explain the main function of each part.

Rana pipiens or *Rana catesbeiana*

Frogs are the most commonly studied representatives of the Class Amphibia, Subphylum Vertebrata, Phylum Chordata (figure 16.1). Although there is no "typical vertebrate" any more than there is a "typical person," a study of frog anatomy does serve to illustrate effectively the basic body organization of a vertebrate animal.

It is important to remember that the amphibians are transitional animals that typically live a portion of their lives on land and another portion in the water. Thus, they exhibit a peculiar mixture of characteristics—some of which represent adaptations for terrestrial life, and some represent adaptations for life in the water. Mating nearly always occurs in the water, since the eggs lack the protective outer coverings that permit the terrestrial existence of birds and reptiles. Amphibian eggs hatch into **tadpole larvae** with gills and a muscular tail for swimming. Later the tadpoles metamorphose into four-legged adults, which are usually semiterrestrial. Some amphibians, however, spend all of their lives in the water, and a very few spend all their lives on land, having developed special mechanisms to protect their eggs from desiccation.

The skin of adult amphibians is smooth and usually moist. Generally the adults live in wet or moist environments, since they are very susceptible to water loss through the skin. Toads are rather exceptional amphibians that have developed a tough, horny skin that reduces water loss and thus allows them to spend more time on land.

Since the amphibians represent an evolutionary transition between the fishes and the terrestrial animals, they also exhibit numerous morphological advances over the fishes. The skull is broad, flat, and lighter in weight; they have jointed tetrapod limbs rather than fins for locomotion; adult amphibians often develop lungs for breathing; a three-chambered heart is formed; and the circulatory system exhibits two distinct circuits, a **systemic division** to supply the body organs and a **pulmonary division** to carry blood to and from the lungs.

Frogs exhibit most of the typical amphibian features, but they also show some features peculiar to their own mode of life. The salamanders more nearly represent the typical features of the Amphibia. Some of the specialized features of the adult frog are: (1) the absence of a tail, (2) the loss of certain skull bones, (3) the posterior attachment of the tongue, (4) the absence of ribs, (5) the lack of a distinct neck, and (6) the powerful and highly developed hind limbs.

eyes

external
naris

tympanic
membrane

Fig. 16.1 *Leopard frog,* Rana pipiens. *(Photograph courtesy of Carolina Biological Supply Company, Burlington, N.C.)*

Materials List
Living specimens
Rana pipiens (or *R. catesbiana*)
Preserved specimens
Rana pipiens (or *R. catesbiana*)
Frog skeleton
Representative vertebrate skeletons

External Anatomy and Behavior

■ Study a living frog and note the smooth, moist, and pliable skin. Observe the pattern of coloration of the dorsal and ventral surfaces of the body.

How does it differ on the two surfaces? Compare the color pattern of your specimen with those of others in the laboratory. How much variation in color patterns do you observe? The specific pattern of spots has been shown to be genetically determined.

Note the broad, flat head with the large mouth, the nostrils or **external nares,** two conspicuous **eyes,** and the circular **tympanic membranes** located behind the eyes. Bordering the eye is a fleshy lower eyelid and a less-prominent upper eyelid. A third transparent inner eyelid, the nictitating membrane, helps keep the eye moist while the frog is on land and also helps protect the eye under water.

■ Observe demonstrations of living frogs in an aquarium and in a terrarium. Study a frog in the aquarium and observe its swimming behavior.

How does the frog propel itself through the water? Which limbs are most prominent in swimming? What adaptations of the limbs are most important for swimming? What is the posture of the frog while it is floating? Compare the behavior of frogs in an aquarium filled with water at room temperature and the behavior of frogs in an aquarium filled with cold water. Which frogs are more active? How can you explain the difference in their activity? Which frogs remain submerged for longer periods of time, those in the cold aquarium or those in the warm (room temperature) aquarium? Frogs are **poikilothermal** or "cold-blooded" animals. This means that their body temperature approximates that of their environment. How does this fact help you to explain the difference in behavior between the frogs in the cold aquarium and those in the warm aquarium?

■ Observe also the behavior of frogs in a terrarium.

When exposed to the air, frogs breathe by drawing air through the external nares into the oral cavity. Later, air is forced into the lungs by closing the internal nares and raising the floor of the mouth. Some gas exchange with the blood also occurs across the moist epithelium of the oral cavity and through the skin covering the exterior of the body.

■ Compare the locomotion of frogs in the water with frogs in a terrarium or on a laboratory table. How are the limbs used in locomotion on land?

To observe the feeding behavior of a frog, place some living house flies, fruit flies, or other small insects in a small aquarium with a frog. How does the frog react to the food organisms? How are the insects captured?

Skeletal System

The skeleton of the frog (figures 16.2 and 16.3) consists chiefly of bone and cartilage. It supports the various parts of the body, protects delicate organs such as the brain and spinal cord, and provides points of attachment for the skeletal muscles.

■ Study a prepared skeleton of a bullfrog (plastic-embedded specimens are excellent for this purpose) and compare the general organization of the frog skeleton to that of other vertebrates on demonstration.

The skeleton of **vertebrates** consists of the **somatic skeleton** (skeleton of the body wall and appendages) and the **visceral skeleton** (skeleton of the pharyngeal wall—prominent in fishes as support for the gills and as a part of the jaws but much reduced in higher vertebrates). In the frog, the visceral skeleton is represented principally by the **hyoid apparatus,** a small bone and cartilage structure that helps support the floor of the mouth, the base of the tongue, and parts of the jaws and larynx.

The major components of the frog skeleton and some of their relationships are shown in the following outline.

> Visceral Skeleton
> > Hyoid apparatus
>
> Somatic Skeleton
>
> > Axial skeleton
> > > Vertebral column
> > > Sternum
> > > Skull
> >
> > Appendicular skeleton
> > > Pectoral girdle
> > > Forelimbs
> > > > Humerus
> > > > Radio-ulna
> > > > Carpals
> > > > Metacarpals
> > > > Phalanges
> > >
> > > Pelvic girdle
> > > Hindlimbs
> > > > Femur
> > > > Tibio-fibula
> > > > Tarsals
> > > > Metatarsals
> > > > Phalanges

The **skull** consists of three main parts (figure 16.2): (1) the narrow braincase or **cranium;** (2) the paired **sensory capsules** of the ears, nose, and the large eye sockets, or orbits, for the eyes, and (3) the **visceral skeleton** (consisting of parts of the jaws, hyoid apparatus, and the laryngeal cartilages).

■ Study the figures and locate the various bones of the skull on your specimen.

The **vertebral column** is made up of ten vertebrae. The first vertebra, the **atlas,** articulates with the base of the skull (see figure 16.2). It has no transverse processes and is the only **cervical** (neck) **vertebra** in the frog. The next seven vertebrae are the **abdominal vertebrae** (figure 16.3). Behind the abdominal vertebrae is a large **sacrum** with two strong transverse processes, which join with the **ileum.** The last vertebra is the long **urostyle** (figure 16.3).

■ Study the parts of an abdominal vertebra with the aid of figure 16.3.

Muscular System (Optional)

Effective study of the muscular system of the frog takes time, patience, and good dissection technique. There are some 200 muscles in the frog, so a complete study of its musculature would take a good deal of time. In this exercise, we shall concentrate only on a few muscles to illustrate the organization of the muscular system and to learn something about the relationship between muscles, bones, and other parts of the body.

Some of the principal muscles of the frog are illustrated in figures 16.4 and 16.5. The skeletal muscles of the adult are well adapted for swimming and for locomotion on land. Note the large, highly developed muscles associated with the hindlimbs.

A skeletal muscle typically consists of a fixed end, called the **origin,** and a movable end, called the **insertion.** The fleshy middle portion is called the **belly.** Most of the skeletal muscles taper at their ends into tough white cords of connective tissue, the **tendons,** which serve to attach them to bones or to other muscles.

The movement or effect caused by a muscle is its **action.** Thus, each skeletal muscle has a characteristic origin, insertion, and action. Most muscles are arranged in pairs or groups that are **antagonistic;** they have opposite actions.

Muscles are classified according to their actions. Some general classes of skeletal muscles are: **extensors,** muscles that straighten or extend a part; **flexors,** muscles that bend one part toward another part; **adductors,** muscles that pull a part back toward the axis of the body; and **abductors,** muscles that draw a part away from the axis of the body.

Fig. 16.2 *Frog skull.*

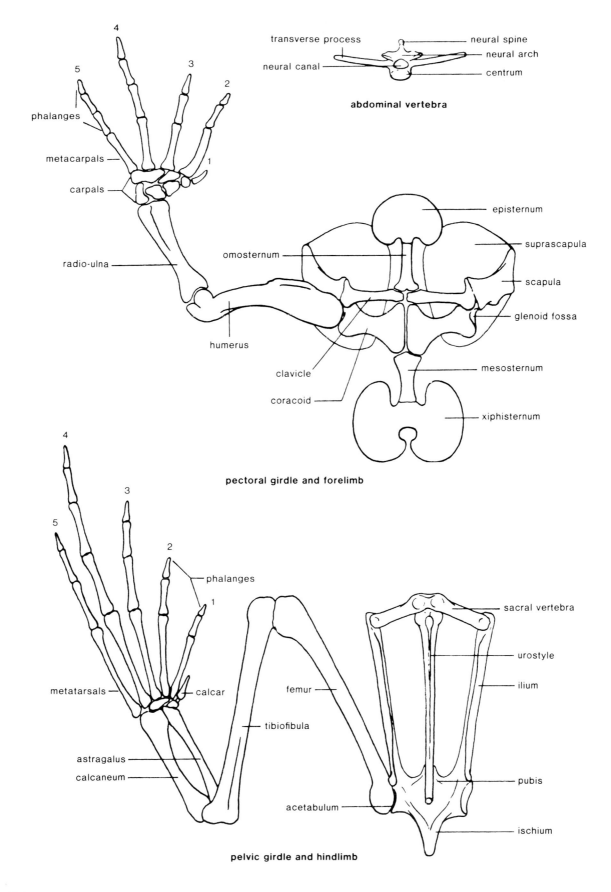

Fig. 16.3 *Frog skeletal system.*

Frog Anatomy

temporal
masseter
cucullaris
deltoid
triceps brachii
anconeus
iliolumbaris
longissimus dorsi
iliacus internus
adductor longus
triceps femoris
adductor magnus
gracilis
major
gracilis
minor
semi-
membranosus
pterygoideus
depressor mandibularis
dorsalis scapulae
extensor carpi-radialis
ulnaris
latissimus dorsi
external oblique
cutaneous abdominis
gluteus
piriformis
triceps femoris
biceps femoris
peroneus
gastrocnemius
tibialis anticus

Fig. 16.4 *Frog. Muscular system, dorsal view.*

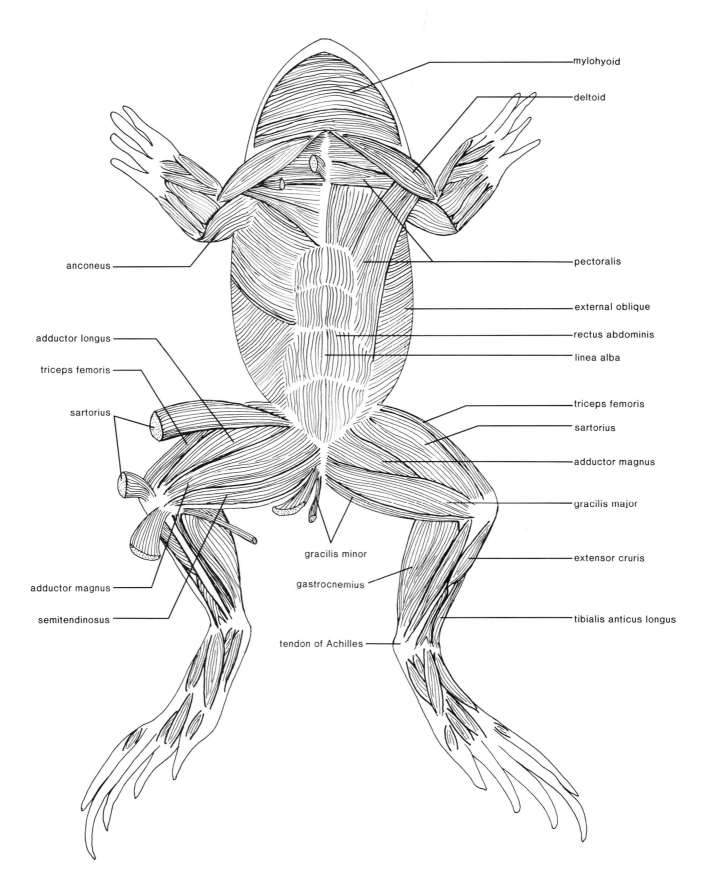

mylohyoid

deltoid

anconeus

pectoralis

external oblique

rectus abdominis

linea alba

adductor longus

triceps femoris

triceps femoris

sartorius

sartorius

adductor magnus

gracilis major

gracilis minor

extensor cruris

adductor magnus

gastrocnemius

semitendinosus

tibialis anticus longus

tendon of Achilles

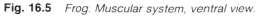

Fig. 16.5 *Frog. Muscular system, ventral view.*

For a detailed study of the muscles of the frog, you must first remove the skin from a specimen. Specimens especially preserved in alcohol are best for this purpose although formalin-preserved specimens may also be used. The chemicals used for preservation can be irritating to the skin and eyes. You should wear plastic or rubber gloves and protective eye gear when dissecting preserved specimens. See the section on laboratory safety following the preface of this book for further information about the properties of chemicals used for preservation and for other safety precautions. If you follow a few simple safety rules, work in the biology lab can be safe and rewarding.

Skinning the Frog

1. Place the frog ventral surface up in a dissecting pan. Carefully cut through the skin along the midventral line from the anterior tip of the jaw to the region of the cloaca. Pull the skin up and away from the underlying muscles to avoid damage to the muscles.

2. Make two transverse cuts through the skin around the body anterior to the forelimbs and anterior to the hind limbs.

3. Make additional cuts around the opening of the cloaca, and around the eyes and tympanic membranes on each side of the head.

4. Carefully loosen the skin from the underlying muscles with a **blunt** (not sharp!) **instrument.** Observe the thin white sheets of connective tissue between the skin and the muscles. Note the large areas where the skin is not attached to the muscles. These areas are **subcutaneous lymph sacs** where this colorless fluid collects in a living frog.

5. Gently peel off the skin to expose the muscles of the abdomen, head, and appendages. Loosen the skin at the base of each limb and carefully pull from the base toward the distal end of the appendage, turning the skin inside out.

After you have finished skinning the frog, compare your specimen with figures 16.4 and 16.5. Identify first the principal muscles of the abdomen. Carefully free and separate the muscles with a blunt instrument as you work. Follow the direction of the muscle fibers as you gently separate one muscle from another.

Muscles of the Hind Limb

A study of the muscles of the well-developed hindlimbs provides an excellent introduction to the principles of muscle anatomy. Table 16.1 lists the principal muscles of the hindlimb. Start with the muscles of the dorsal surface of the thigh (upper leg) and identify the five principal muscles on this surface. Observe how these muscles provide control of the various movements of the leg and foot.

Continue your study with the muscles of the ventral surface of the thigh and then identify the muscles of the shank (lower leg). How many pairs of antagonistic muscles can you identify from the hindlimb?

Oral Cavity

Obtain an anaesthetized frog from your instructor to study the anatomy of the **oral cavity.**

Frogs can be anaesthetized by immersion in a solution of methyl sulphonate or urethane. You and your instructor should make sure that your frog is properly anaesthetized before you proceed.

Place the frog on its back and make a small cut with your scissors at each corner of the mouth so the jaws can be opened widely. Rinse out the oral cavity with cold water if there is an accumulation of mucus. Consult figure 16.6 and locate the following structures in the oral cavity of your specimen: the **maxillary teeth** on the margin of the upper jaw, the **vomerine teeth** on the palate (the teeth are used for holding food rather than for chewing since the food is swallowed whole); the **internal nares** (probe through them from the **external nares**); and the openings into the **eustachian tubes.** Observe also the **tongue** and its attachment.

What advantage does this peculiar attachment of the tongue have for the frog?

Posterior to the oral cavity (behind the tongue) is the **pharynx.** At the rear of the pharynx is the opening into the esophagus, and on the ventral surface find the **glottis,** a slitlike opening into the **larynx.** A pair of short **bronchi** connect the larynx to the two lungs.

The males of many species of frog have openings into two **vocal sacs** located at the rear of the mouth just anterior to the openings to the eustachian tubes. Inflation of the vocal sacs serves to amplify the croaking sounds. Vocal sacs are absent in female frogs.

Internal Anatomy

Before continuing with your study of internal anatomy of the frog, you should review the general instructions for dissection provided in chapter 9.

If you are dissecting an anaesthetized specimen, take care to minimize the cutting of blood vessels to prevent the blood from obscuring your view of the internal organs. Keep your specimen covered with cold water during dissection. Fasten the frog, ventral side up, to the bottom of the dissection pan with pins through the tip of the jaw and all the limbs.

If you have a preserved specimen for dissection, it may be saturated with the preservation fluid or it may simply be moist and come in a plastic bag. In either case, the tissues should be moist and pliable.

Study the general location of internal organs of the frog in figures 16.7 and 16.8 to make certain that you understand the approximate location of the internal

Table 16.1 Chief Muscles of the Thigh and Shank

Muscle	Location	Origin	Insertion	Action
Dorsal Muscles of the Thigh				
Triceps femoris (three parts)	Lateral surface of thigh	Ilium, acetabulum	Tibiofibula	Flexes thigh and extends shank
Rectus anterior femoris		Ileum	Fascia attached to tibiofibula	Flexes thigh and extends shank
Vastus internus		Acetabulum	Tibiofibula	Flexes thigh and extends shank
Vastus externus		Ileum	Vastus internus	Flexes thigh and extends shank
Gluteus	Medial and anterior to triceps femoris	Ileum	Femur	Rotates thigh
Semimembranosus	Medial to triceps femoris	Ischium and pubis	Tibiofibula	Extends thigh and flexes shank
Biceps femoris (iliofibularis)	Between triceps femoris and semimembranosus	Ileum	Tibiofibula and femur	Extends and adducts thigh; flexes shank
Piriformis	Small muscle near cloacal opening; between biceps femoris and semimembranosus	Urostyle	Femur	Extends and rotates thigh
Ventral Muscles of the Thigh				
Sartorius	Large, flat midventral muscle	Pubis	Tibiofibula	Flexes thigh and shank
Gracilis major (rectus internus major)	Medial to sartorius	Pubis	Tibiofibula	Extends thigh and flexes shank
Gracilis minor (rectus internus minor)	Medial to gracilis major	Pubis	Tibiofibula	Extends thigh and flexes shank
Adductor magnus	Medial and partly beneath sartorius	Ischium and pubis	Femur	Adducts thigh
Adductor longus	Beneath sartorius	Pubis	Femur	Adducts thigh
Semitendenosus	Under and between gracilis minor and adductor magnus	Ischium	Tibiofibula	Extends and adducts thigh, flexes knee
Muscles of the Shank				
Gastrocnemius	Medial surface of shank	Femur	Achilles tendon	Flexes shank and foot
Peroneus	Lateral to gastrocnemius	Femur	Distal end of tibiofibula	Extends shank and foot
Tibialis anticus longus (tibialis anterior longus)	Anterior to tibiofibula	Femur	Tarsal bones	Extends foot and flexes ankle
Tibialis posticus (tibialis posterior)	Posterior to tibiofibula	Tibiofibula	Tarsal bones	Flexes foot
Extensor cruris	Anterior to and partly beneath tibialis anticus longus	Femur	Tibiofibula	Extends shank

organs before you begin your dissection. Proceed carefully in your dissection to avoid unnecessary damage to body parts that you may need to study later.

Lift the skin from the body with your forceps and make a **longitudinal cut** slightly to one side of the midventral line and forward from the pelvis to the tip of the lower jaw. Observe the subcutaneous lymph spaces and the attachment of the skin to the body as you free the skin from the ventral body surface. Make short transverse cuts in the skin at the anterior and posterior ends of the trunk and pin back the flaps of skin on both sides.

Consult figure 16.5 and locate the following three large muscles before continuing with your dissection: the **pectoralis**, the **rectus abdominis**, and the **external oblique**.

internal
nares

vomerine
teeth

openings
to
eustachian
tubes

tongue

Fig. 16.6 *Frog. Oral cavity, ventral view. (Photograph courtesy of Carolina Biological Supply Company, Burlington, N.C.)*

Note also the whitish **linea alba** along the midventral line. The ventral abdominal vein lies beneath the muscular body wall along the midventral line.

Now lift the muscles of the abdomen with your forceps and make a **longitudinal incision** through the body wall with your scissors a little to one side of the linea alba. Cut through the bony **sternum** which supports the forelimbs (see figure 16.3), and up to the posterior end of the lower jaw, taking care not to cut the ventral abdominal vein and to avoid damage to the internal organs within the coelom. Trace the ventral abdominal vein to the liver and make two short transverse incisions through the body wall just in front of the liver so the muscles can be pinned back to expose the internal organs. Remove about a half-inch section from the middle portion of the sternum and pin the forelimbs back to provide access to the internal organs in this region as shown in figure 16.7.

Survey of Internal Organs

Now that you have completed the preliminary phase of your dissection, make a brief survey of the internal anatomy, referring to the figures cited in the following description. Note the position, size, shape, color, and texture of each organ. Consider also the relationship of each organ to other organs and attempt to determine the function(s) of each organ as you work. If you have a female specimen, and the ovaries and oviducts are filled with eggs and greatly enlarged, consult figure 16.14 and carefully remove the ovary and oviduct from one side of the body to facilitate your study of the other parts.

The location of the principal internal organs is illustrated in figures 16.7 and 16.8. The large, three-lobed **liver,** located just posterior to the pectoral girdle, is one of the most prominent internal organs. The liver is dark in color in both preserved and living specimens. Adjacent to the liver on the right side of the frog (viewed from the ventral surface as in a dissecting pan) is the curved, tubular **stomach.**

The **heart** is enclosed in a membranous sac, the **pericardial cavity,** which lies anterior to the liver and partly beneath the pectoral girdle. You will open the pericardial cavity later to study the heart and its relationship with its attached blood vessels. The two **lungs** are located on the sides of the visceral cavity posterior and lateral to the heart and dorsal to the liver. In a living or freshly killed frog, the lungs will be semitransparent, pink, and inflated. In preserved specimens, they will often be deflated.

If your specimen is a mature female, much of the visceral cavity will be filled with many black and white **eggs** enclosed in the large, membranous **ovaries.**

After you have located these major interior organs, you should proceed with a more detailed study of specific organ systems as directed by your laboratory instructor and as described in the following sections.

Digestive System

Trace the pathway of the digestive system starting with the mouth and the **oral cavity.** Behind the **tongue,** locate the **pharynx** and the opening into the **esophagus.** The esophagus is a short, cylindrical tube that passes food material to the **stomach,** where the food is stored temporarily and digestion begins. The muscular stomach also provides a thorough mixing of the food mass by its contractions. At the posterior end of the stomach, find the **pyloric valve,** a sphincter that controls the release of the stomach contents into the small intestine.

The anterior segment of the small intestine is the **duodenum,** which receives secretions from the liver and pancreas through the common bile duct. Behind the duodenum is the convoluted **ileum,** the posterior section of the small intestine where digestion is completed and where most absorption of nutrients into the bloodstream occurs. The ileum empties into the **large intestine,** where most water as well as certain vitamins and ions are absorbed.

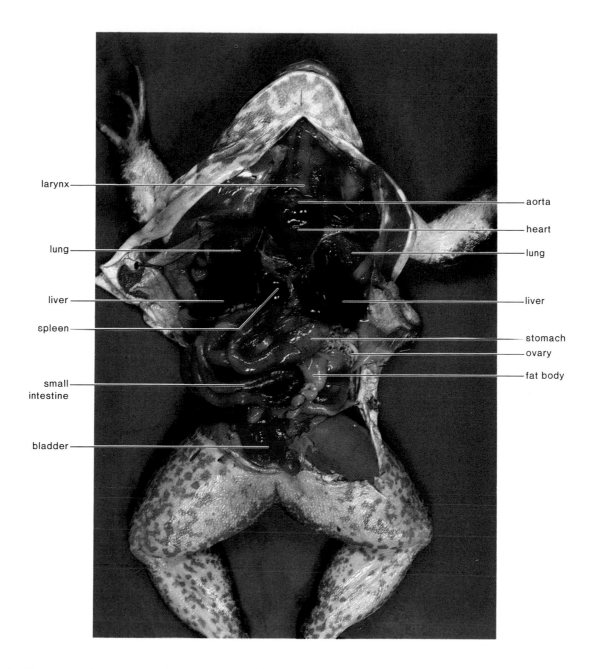

larynx

lung

liver

spleen

small intestine

bladder

aorta

heart

lung

liver

stomach

ovary

fat body

Fig. 16.7 *Frog. Internal organs, female, ventral view. (Photograph by Ken Taylor.)*

The large intestine is also where the undigested residue is temporarily stored as fecal material. Posteriorly the large intestine narrows and empties into the **cloaca.** The cloaca is a common chamber that collects materials from the digestive, excretory, and reproductive systems prior to their discharge through the cloacal opening.

Respiratory System

Adult frogs accomplish the necessary exchange of gases by using three different parts of their body for respiration: the **skin,** the **lungs,** and the **lining** of the **oral cavity** and **pharynx.** Locate the following parts of the respiratory system of the frog, which supplement the respiration accomplished through the skin.

Locate the **external nares** on the anterior part of the head (figure 16.1) and the **internal nares** on the roof of the oral cavity (figure 16.6) where air enters. The **nasal cavity** connects the internal and external nares on each side. From the **pharynx,** posterior to the oral cavity, air passes through the **glottis** into the **larynx,** a cartilaginous tube that divides posteriorly to form two **bronchial tubes** or **bronchi.** The two bronchial tubes connect with the two **lungs.** The lungs are located dorsal of the heart and liver, one on each side of the body (figure 16.7). Inside each lung are thousands of tiny **air sacs,** each surrounded by numerous capillaries. In these sacs, the exchange of gases with the blood takes place.

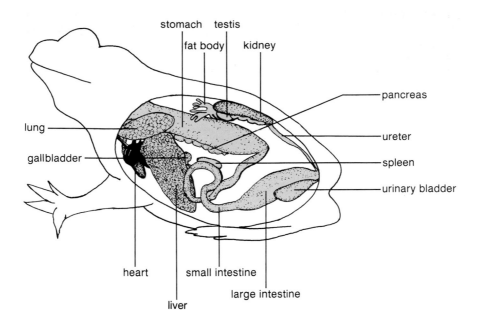

stomach testis
fat body kidney
pancreas
lung
ureter
gallbladder
spleen
urinary bladder
heart small intestine
large intestine
liver

Fig. 16.8 *Frog. Internal organs, lateral view.*

Circulatory System

Amphibians have a **three-chambered heart** with two **atria** and a **ventricle** (figure 16.7). In the adult frog, part of the venous blood moves from the heart to the respiratory organs (lungs) and then returns to the heart before it is pumped out to the various organs of the body. Thus there are two major divisions of the circulatory system: (1) the **pulmocutaneous circulation** and (2) the **systemic circulation**. Preserved frogs that have been injected with latex are most satisfactory for detailed study of the circulatory system. Injected specimens may be singly injected (arteries only) or doubly injected (both arteries and veins injected with latex). Arteries usually are filled with red latex and veins are filled with blue latex.

Heart. Carefully remove the enveloping membrane, the **pericardium**, from around the heart. Consult figures 16.7 and 16.9, and identify the thin-walled right and left **atria**, the thick-walled **ventricle**, and the **conus arteriosus**, which divides into two branches, the right and left **truncus arteriosus**. Lift the ventricle and find the dark-colored, thin-walled, triangular **sinus venosus** on its dorsal side (figure 16.12), the **posterior vena cava** entering it at the posterior end, and the right and left **anterior vena cavae** entering it at the front.

Arterial System. Study the arterial system with the aid of figures 16.9, 16.10, and 16.11. Note the left and right branches of the **truncus arteriosus,** each giving rise to three large arteries called aortic arches.

1. The **common carotid artery.** This divides into: (a) the **external carotid** or lingual artery leading to the ventral part of the head and to the tongue; (b) the **internal carotid** to the dorsal part of the head. (Note the carotid gland.)

2. The **systemic arch.** The two systemic arches bend dorsally around the pharynx and unite to form the large **dorsal aorta.** Note the arteries leading from each systemic arch:
   a. the short **occipito-vertebral** dividing into the **occipital** leading to the skull and the **vertebral** leading to the vertebral column;
   b. the **subclavian,** giving off branches to the shoulder region and extending into the arm where it becomes the brachial;
   c. the **dorsal aorta;**
   d. the **coeliacomesenteric,** which divides into the **coeliac** and the **mesenteric;**
   e. the **right** and **left gastric** to the stomach, the **pancreatic** to the pancreas, and the **hepatic artery** to the liver (these latter arteries are all branches of the coeliac);
   f. the **mesenteric** to the intestine, with branches to the spleen (the **splenic artery**) and to the rectum and large intestine (**posterior mesenteric artery**);
   g. the **urogenital arteries** to the kidneys, gonads, and fat bodies (**renal** to kidneys and **genital** to gonads). (The urogenital arteries vary considerably in number and arrangement in different individuals. Sometimes they do not inject well because of their small size and may be difficult to find. How many are there in your specimen?);
   h. the **lumbar arteries** to the dorsal body wall (several pairs of small arteries arising from the dorsal surface of the aorta—not shown in the figure, also small and often hard to find);

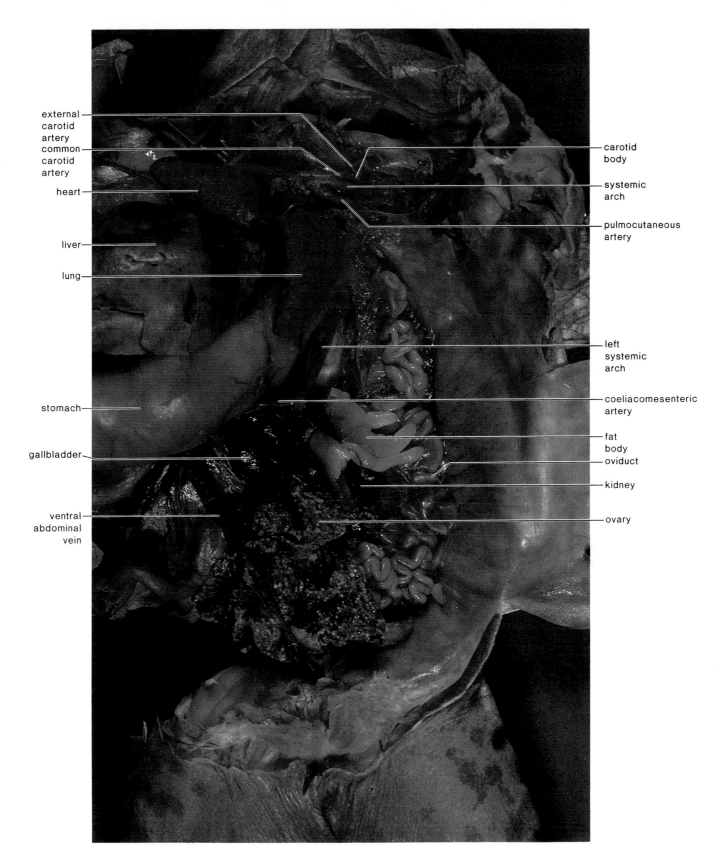

Fig. 16.9 *Frog. Heart and major arteries, ventral view. (Photograph by Ken Taylor.)*

external
carotid
artery

common
carotid
artery

heart

liver

lung

stomach

gallbladder

ventral
abdominal
vein

carotid
body

systemic
arch

pulmocutaneous
artery

left
systemic
arch

coeliacomesenteric
artery

fat
body

oviduct

kidney

ovary

lung

stomach

celiacomesenteric
artery

urogenital
arteries

ovary

kidney

lumbar
arteries

dorsal
aorta

common
iliac
artery

Fig. 16.10   *Frog. Dorsal aorta and abdominal arteries, ventral view. (Photograph by Ken Taylor.)*

i.   the **common iliac arteries,** divisions of the dorsal aorta;

j.   the **epigastric** to the bladder and body wall of that region;

k.   the **femoral** in the thigh;

l.   the **sciatic,** a continuation of the iliac furnishing branches to most of the muscles of the leg; and

m.   the **peroneal** and **tibial arteries,** divisions of the sciatic to lower parts of the leg as shown in the figure.

3.   The **pulmocutaneous artery.** This divides into (a) the **pulmonary artery** going to the lung and (b) the **cutaneous artery** to the skin. Note that this artery and its branches carry nonoxygenated blood.

Venous System. Veins are blood vessels that carry blood **toward** the heart. Study the venous system of your specimen with the aid of figure 16.12 and locate the following veins.

1.   The two (right and left) **anterior venae cavae** into which enter: (a) the **external jugular** with its branches, the **lingual** from the tongue and floor of

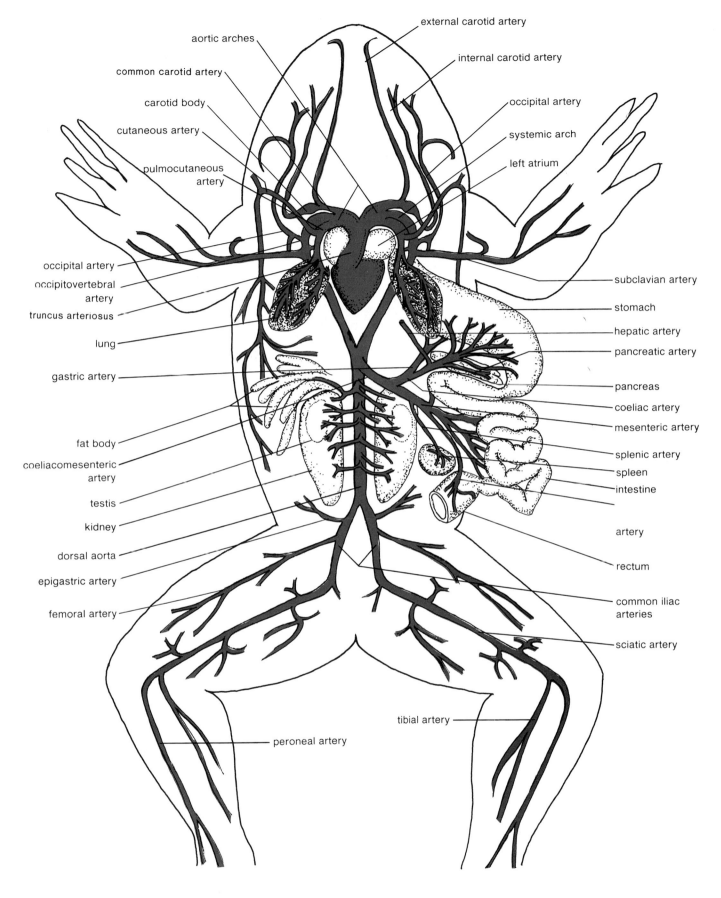

Fig. 16.11 *Frog. Arterial system, ventral view.*

Frog Anatomy

239

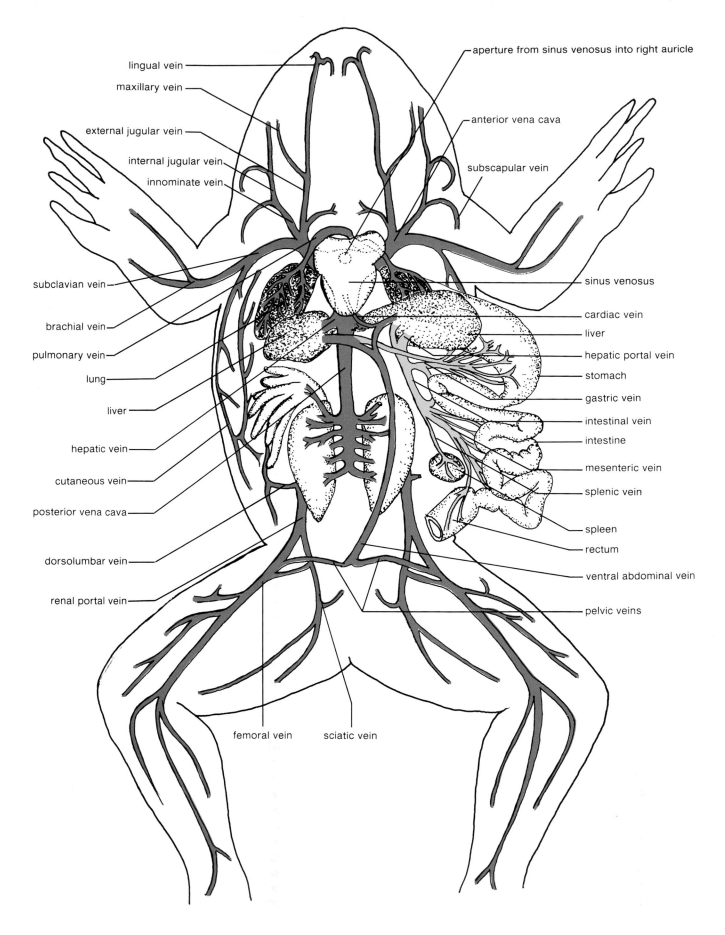

Fig. 16.12 *Frog. Venous system, ventral view.*

Labels (left side, top to bottom):
lingual vein
maxillary vein
external jugular vein
internal jugular vein
innominate vein
subclavian vein
brachial vein
pulmonary vein
lung
liver
hepatic vein
cutaneous vein
posterior vena cava
dorsolumbar vein
renal portal vein

Labels (right side, top to bottom):
aperture from sinus venosus into right auricle
anterior vena cava
subscapular vein
sinus venosus
cardiac vein
liver
hepatic portal vein
stomach
gastric vein
intestinal vein
intestine
mesenteric vein
splenic vein
spleen
rectum
ventral abdominal vein
pelvic veins

Labels (bottom):
femoral vein
sciatic vein

the mouth and **maxillary** or mandibular from the jaw; (b) the **innominate vein** into which flows the **internal jugular** from the deeper parts of the head and the **subscapular** from the shoulder; and (c) the **subclavian vein** formed by the union of the **musculocutaneous** from the muscles and skin of the side and back, and the **brachial** from the forelimb.

2. The **posterior vena cava** into which enter: (a) the **hepatic veins** from the liver; (b) the **renal veins** from the kidneys; and (c) the **genital veins** from the gonads.

3. The **hepatic portal system** consists of: (a) the **abdominal vein,** which enters the liver and is formed by the union of the two **pelvic veins;** (b) the **hepatic portal vein,** which carries blood from the stomach (**gastric vein**), intestine (**mesenteric vein**), and the spleen (**splenic vein**).

4. The **renal portal system.** The **renal portal veins** carry the blood to the kidneys. They receive the blood from (a) the **dorso-lumbar vein;** (b) a branch of the **femoral vein** from the hindlimb; and (c) the **sciatic vein** from the thigh. Note that the blood from the femoral vein may go to the kidney by way of the renal portal vein or to the liver by way of the pelvic and abdominal veins.

5. The **pulmonary veins.** These veins carry oxygenated blood from the lungs and unite to enter the left auricle.

The Urogenital System

The excretory and reproductive organs are closely associated, and together comprise the **urogenital system.** If your specimen is a female with large ovaries concealing the other visceral organs, consult figures 16.13, 16.14, and 16.15, and carefully remove one of the ovaries. The excretory structures are similar in the two sexes of the frog. With the help of figure 16.13, again note the two **kidneys,** each with the adrenal gland on its ventral surface; the **ureter** leading from the posterior border of the kidney to the cloaca; the **urinary bladder,** which empties into the cloaca on the ventral side; and the **fat bodies.**

The kidneys are the major excretory organs of the frog. They are responsible for the removal of most of the wastes from the body, although some wastes are also lost through the skin. The kidneys remove most of the nitrogenous wastes from the blood, which are excreted in soluble form as urea and ammonia. These important organs also play a major role in **homeostasis,** the maintenance of a constant internal environment by removing excess ions and other substances from the blood and by conserving those substances of limited availability.

Specimens should be distributed in the class so that both male and female specimens are available at each table. After you have completed the study of the urogenital system on your own specimen, find a specimen of the opposite sex and make a comparative study so that you will be familiar with the organization of the urogenital system in both the male and female.

In a female specimen (figures 16.13 and 16.14), locate on one side the **ovary** attached dorsally and suspended into the coelom by a **mesentery;** the coiled **oviduct** with a funnel-shaped opening, the **ostium,** at its anterior end; and a **uterus,** an enlargement near the cloaca. Cut open the cloaca and use your probe to find the openings of the oviducts, the ureters, and the urinary bladder.

Near the time of maturation of the eggs, the eggs are released from the ovaries into the coelom and pass through the ostia into the oviducts. As the eggs pass down the oviduct, they are covered with several layers of jellylike material secreted by glands in the walls of the oviducts. The eggs collect in the uteri before they are discharged through the cloaca and into the water. Fertilization is external and in mating the male mounts the female and discharges sperm on the eggs as they are discharged.

In a male specimen (figures 16.13 and 16.15), locate the two **testes,** each suspended from the dorsal abdominal wall by a mesentery, and the several **vasa efferentia,** the small ducts that carry sperm from the testes to the kidneys. The sperm are carried from the kidneys to the cloaca via the **ureters,** which serve as genital ducts in the male. Male leopard frogs also frequently have vestigial oviducts (mesonephric ducts) alongside the kidneys.

Nervous System

The nervous system of vertebrates consists of three main parts: (1) the **central nervous system,** which includes the brain and the spinal cord; (2) the **peripheral nervous system,** which includes the nerves extending from the central nervous system; and (3) the **autonomic nervous system,** a specialized portion of the peripheral nervous system that regulates the glands of the body and the visceral organs.

■ Successful study of the nervous system of the frog requires careful dissection. Do not rush the dissection, because you may damage important structures necessary for your study. Carefully remove the skin from the dorsal surface of the head between the eyes and along the vertebral column. Clear away the muscles and connective tissue underlying the skin to expose the skull and vertebral column. With your scalpel, carefully shave thin sections of bone from the skull until you expose the brain. Then carefully pick away additional small pieces of bone with your forceps until the brain is completely exposed. Continue the same procedure to expose the vertebral column and the spinal cord, but leave the nervous system in place.

The Brain. Consult figures 16.16 and 16.17 and identify the five main regions of the brain: the **telencephalon,** the **diencephalon,** the **mesencephalon,** the **metencephalon,** and the **myelencephalon.** The anterior-most telencephalon bears four distinct lobes: two **olfactory lobes** and, posterior to them, two **cerebral hemispheres.** Extending anteriorly from the olfactory lobes are the **olfactory nerves,** which carry impulses from the olfactory epithelium to the brain.

241

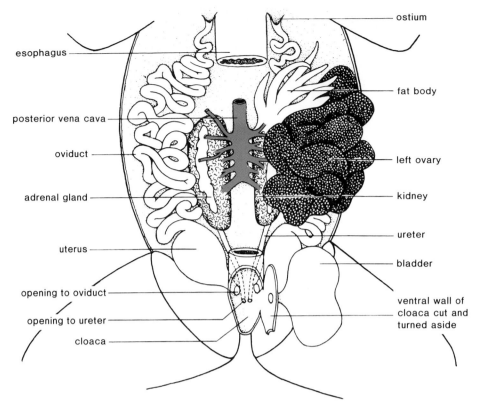

esophagus

ostium

posterior vena cava

fat body

oviduct

left ovary

adrenal gland

kidney

uterus

ureter

bladder

opening to oviduct

opening to ureter

ventral wall of
cloaca cut and
turned aside

cloaca

female

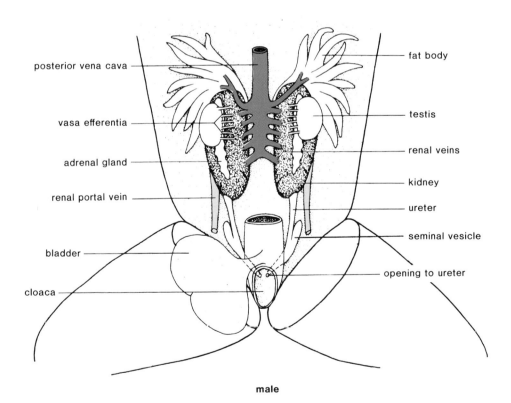

posterior vena cava

fat body

vasa efferentia

testis

adrenal gland

renal veins

renal portal vein

kidney

ureter

bladder

seminal vesicle

cloaca

opening to ureter

male

Fig. 16.13 *Frog. Urogenital system, male and female, ventral view. Systemic veins are blue, portal veins are yellow.*

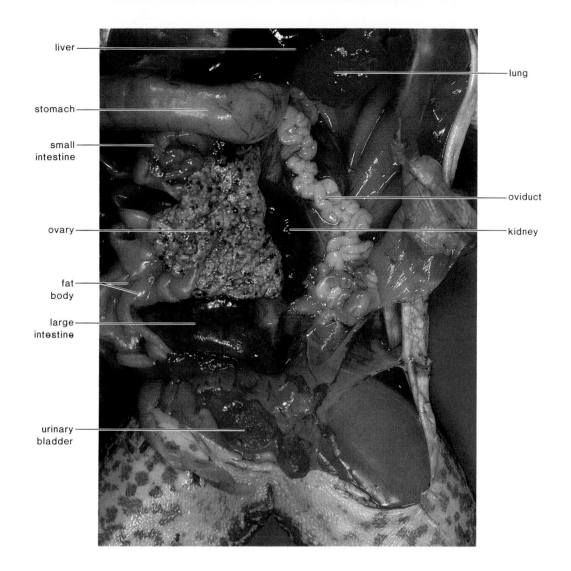

liver
stomach
small intestine
ovary
fat body
large intestine
urinary bladder
lung
oviduct
kidney

Fig. 16.14 *Frog. Female reproductive organs, ventral view. (Photograph by Ken Taylor.)*

Posterior to the telencephalon is the diencephalon, a diamond-shaped depressed area directly behind and somewhat between the posterior portion of the cerebral hemispheres. The **pineal gland** (epiphysis) is a small, inconspicuous body attached to the dorsal wall of the diencephalon, which probably will be removed in your dissection of the skull. The stalk by which it was attached, however, may still be visible. The lateral walls of the diencephalon constitute the **thalamus.** Below the diencephalon lies the **pituitary gland,** which you will study later. The principal function of the telencephalon and diencephalon is the integration of olfactory signals. Chemical perception is of great importance to the frog in feeding and in defense.

The mesencephalon, immediately posterior to the diencephalon, bears two large **optic lobes** that serve to integrate nerve impulses from the eyes. The optic lobes in the frog also provide an important overall coordination of sensory information, a function carried out mainly by the cerebral hemispheres in higher vertebrates.

Behind the mesencephalon find the metencephalon, which is represented by the **cerebellum,** a narrow transverse portion of the brain lying immediately posterior to the optic lobes.

The most posterior part of the brain is the myelencephalon, consisting of the elongated **medulla oblongata,** which tapers gradually into the **spinal cord.** The anterior portion of the medulla oblongata has a thin roof that is frequently removed during dissection. The cerebellum and medulla receive and integrate signals from the ears and from the skeletal muscles. They serve to monitor the general state of activity in the body and also coordinate respiration, feeding, and muscular activity.

■ When you have completed your study of the dorsal side of the brain, **carefully** detach the anterior portion of the brain and lift it gently from the floor of the skull. Continue detaching the brain and spinal cord from its ventral connections and gradually free

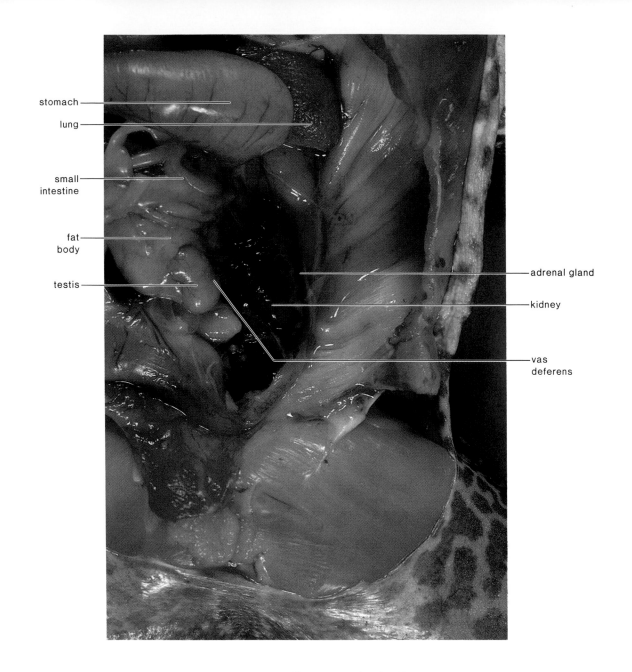

stomach

lung

small intestine

fat body

testis

adrenal gland

kidney

vas deferens

Fig. 16.15 *Frog. Male reproductive system, ventral view. (Photograph by Ken Taylor.)*

the entire nervous system. Place the isolated nervous system ventral side up in your dissecting pan and study the structures visible from this aspect.

Note the crossed optic nerves, called the **optic chiasma,** on the ventral side of the diencephalon. Observe how the nerves from the left eye carry their impulses across the brain to the right optic lobe. The optic nerves are the

second pair of **cranial nerves** extending from the brain to various parts of the body. The olfactory nerves extending forward from the olfactory lobes are the **first pair** of **cranial nerves.**

Cranial Nerves. There is a total of ten pairs of cranial nerves in the frog, extending from the brain to various parts of the body. (Fishes and amphibians generally have ten

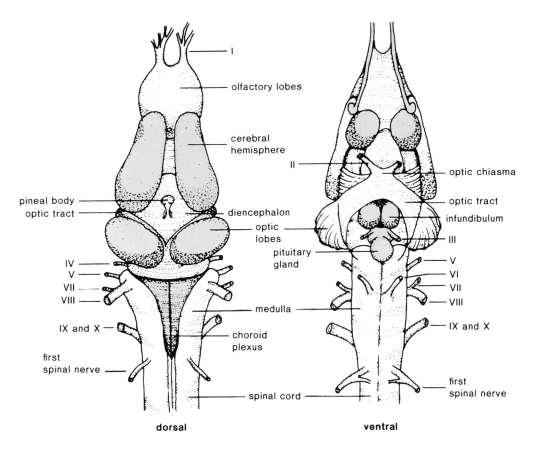

Fig. 16.16 *Frog. Brain and cranial nerves.*

distinct pairs of cranial nerves; reptiles, birds, and mammals have twelve pairs.) Refer to figure 16.16 and locate the following cranial nerves.

Nerve I—Olfactory, extends from the nose to the olfactory lobe.

Nerve II—Optic, extends from the eye to the ventral side of the diencephalon. Note the optic chiasma; see figure 16.16.

Nerve III—Oculomotor, extends from the ventral side of the mesencephalon to the muscles of the eye.

Nerve IV—Trochlear, extends from the dorsal side of the mesencephalon to the superior oblique muscles of the eye.

Nerve V—Trigeminal, closely associated with Nerve VII, the facial, from the side of the medulla to the skin of the face, muscles of the jaw, and tongue.

Nerve VI—Abducens, extends from the ventral surface of the medulla to the external rectus muscle of the eye.

Nerve VII—Facial, mentioned in connection with Nerve V above.

Nerve VIII—Auditory (statoacoustic), extends from the side of the medulla to the ear.

Nerve IX—Glossopharyngeal, extends from the side of the medulla to the muscles and membranes of the tongue and pharynx.

Nerve X—Vagus, closely associated with Nerve IX, extends from the side of the medulla to the heart, lungs, and digestive organs.

Next, observe the ventral outgrowth from the diencephalon, the **infundibulum,** which corresponds to the posterior lobe of the pituitary gland in birds and mammals. Located behind and close to the infundibulum is a dorsal outgrowth originating from the roof of the mouth, the **anterior lobe of the pituitary** (also called the hypophysis). Together these two lobes constitute the **pituitary gland,** an important gland of internal secretion (endocrine gland).

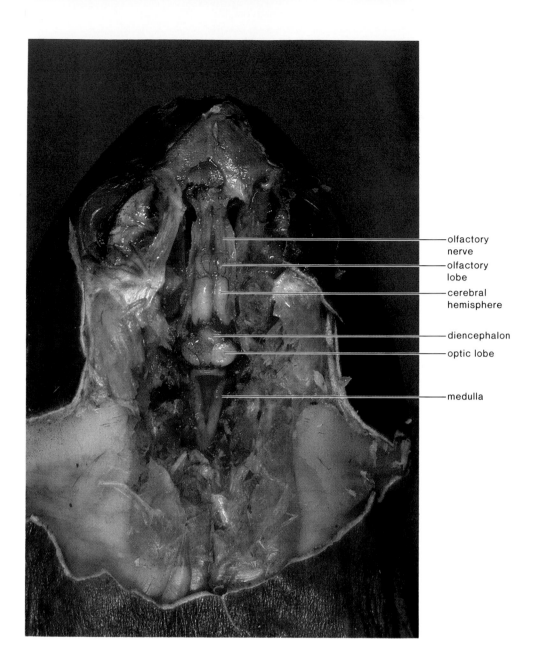

olfactory
nerve

olfactory
lobe

cerebral
hemisphere

diencephalon

optic lobe

medulla

Fig. 16.17 *Frog. Brain, dorsal view. (Photograph by Ken Taylor.)*

Note that the adult pituitary gland develops from two different sources—a ventral outgrowth from the diencephalon (the infundibulum) and a dorsal outgrowth from the mouth (the hypophysis). The hypophysis frequently adheres to the floor of the skull and may be detached during your removal of the brain. If you do not find it on the isolated nervous system, search on the floor of the skull in the appropriate location.

Spinal Cord and Spinal Nerves. Observe now the spinal cord. Note that the terminal portion of the spinal cord in the adult frog is threadlike and is called the **filum terminale.** Ten pairs of **spinal nerves** are attached to the spinal

cord (see figure 16.18). Each nerve has two roots, the **dorsal root** and the **ventral root.** Be careful not to disturb the sympathetic nerves while tracing the spinal nerves with which they unite. Note the large size of the **second spinal nerve.** Find the **brachial plexus** and the **sciatic plexus** where several of the spinal nerves leading to the forelimbs and hindlimbs extend from the spinal cord separately, and later unite and then rebranch. Trace the large **sciatic nerve** to the sciatic plexus and follow its principal branches into the hindlimb.

Autonomic Nervous System. Examine figure 16.19, which shows the autonomic system in solid black. Note the two **sympathetic nerve trunks,** which originate anteriorly as the

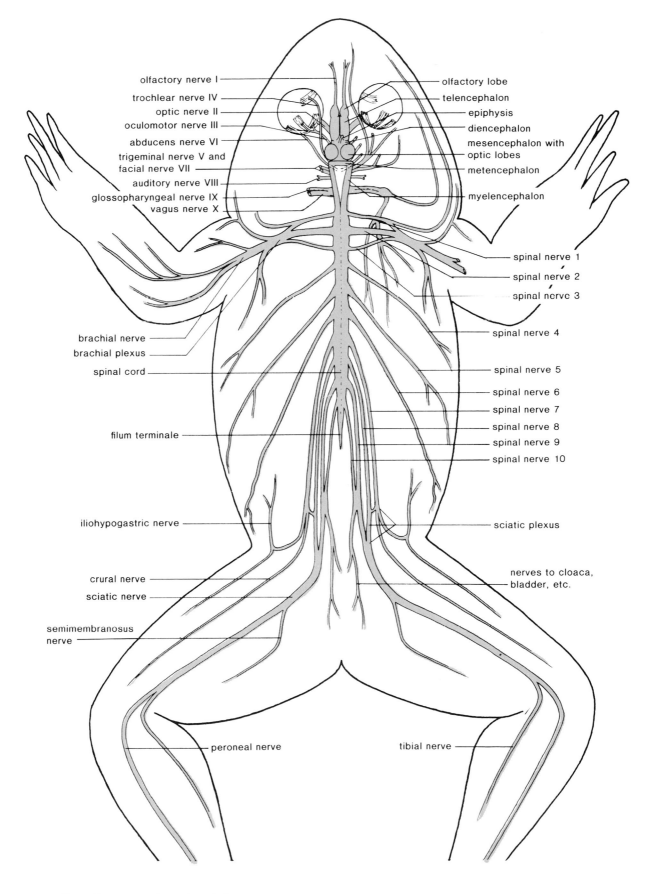

olfactory nerve I
trochlear nerve IV
optic nerve II
oculomotor nerve III
abducens nerve VI
trigeminal nerve V and
facial nerve VII
auditory nerve VIII
glossopharyngeal nerve IX
vagus nerve X

olfactory lobe
telencephalon
epiphysis
diencephalon
mesencephalon with
optic lobes
metencephalon
myelencephalon

spinal nerve 1
spinal nerve 2
spinal nerve 3
spinal nerve 4
spinal nerve 5
spinal nerve 6
spinal nerve 7
spinal nerve 8
spinal nerve 9
spinal nerve 10

brachial nerve
brachial plexus
spinal cord

filum terminale

iliohypogastric nerve

sciatic plexus

crural nerve
sciatic nerve

nerves to cloaca,
bladder, etc.

semimembranosus
nerve

peroneal nerve

tibial nerve

Fig. 16.18 *Frog. Nervous system, dorsal view.*

Frog Anatomy

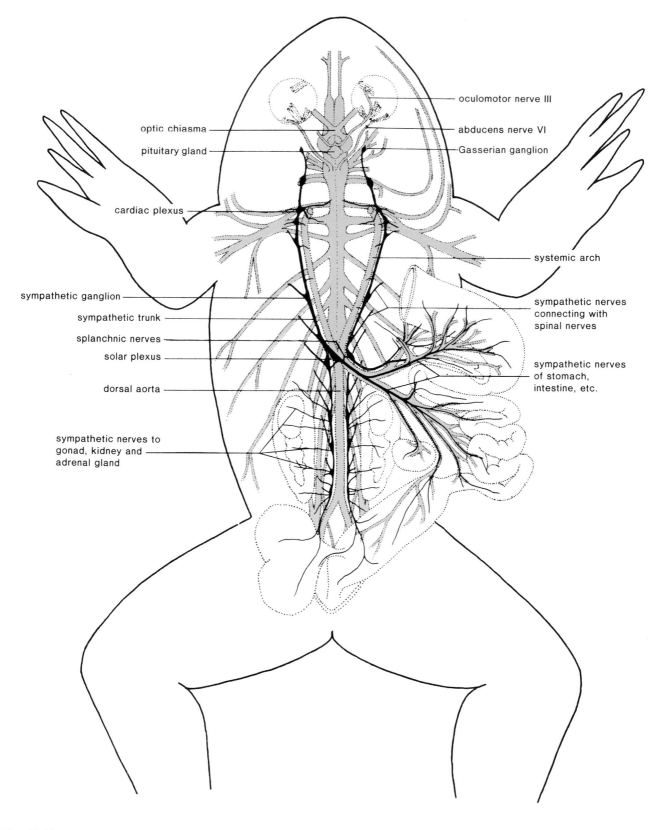

oculomotor nerve III

optic chiasma

abducens nerve VI

pituitary gland

Gasserian ganglion

cardiac plexus

systemic arch

sympathetic ganglion

sympathetic nerves connecting with spinal nerves

sympathetic trunk

splanchnic nerves

solar plexus

sympathetic nerves of stomach, intestine, etc.

dorsal aorta

sympathetic nerves to gonad, kidney and adrenal gland

Fig. 16.19 *Frog. Nervous system, ventral view.*

Gasserian ganglia and extend backward along the systemic arches and dorsal aorta. Find the **sympathetic ganglia,** the **sympathetic nerves** connecting with the spinal nerves, the **cardiac plexus,** the **solar plexus,** the large peripheral **splanchnic nerves** with branches to the stomach, intestines, and other organs, and the delicate **peripheral nerves,** which pass to the gonads, kidneys, spleen, adrenal glands, and other organs.

Demonstrations

1. Bullfrogs—injected and dissected to show the various systems.
2. Various vertebrate skeletons for comparison with the bullfrog.
3. Beating of the frog's heart and circulation of the blood through capillaries of the web of the frog's foot.

Key Terms

Abductor muscle a skeletal muscle that pulls an appendage or body part away from the central axis of the body.

Adductor muscle a skeletal muscle that pulls an appendage or body part toward the central axis of the body.

Autonomic nervous system division of the vertebrate nervous system that controls involuntary functions, such as breathing or heart rate. Consists of branches from several cranial and spinal nerves leading to various internal organs.

Central nervous system portion of the vertebrate nervous system consisting of the brain and the spinal cord.

Cranium portion of the skull that encloses the brain; the braincase.

Extensor muscle a skeletal muscle that extends or straightens an appendage.

Flexor muscle a skeletal muscle that bends an appendage.

Peripheral nervous system portion of the nervous system outside the central nervous system made up of the nerves connecting the organs and tissues with the central nervous system.

Poikilothermal refers to an animal whose body temperature roughly parallels that of his environment; cold-blooded.

Tadpole larva larval form typical of amphibians with an ovoid body and a thin muscular tail.

Urogenital system organ system of vertebrates specialized for excretion and reproduction.

Vertebra unit of the spinal (vertebral) column or "backbone" of the frog and other vertebrates.

Visceral skeleton skeletal parts associated with the support of the gills in fishes and other primitive vertebrates; vestiges remain in the frog and higher vertebrates mainly in the hyoid apparatus that helps support the tongue and trachea.

17

Rat Anatomy

Objectives

1. Locate and identify the principal external features of the rat. Explain which external features distinguish it as a mammal.
2. Identify the main elements of the axial and appendicular skeletons of the rat.
3. Locate five major skeletal muscles of the rat and identify their origin, insertion, and action. Distinguish between extension, flexion, adduction, and abduction.
4. Locate and identify the principal organs of the digestive system of the rat.
5. Explain the general pattern of circulation in the rat and other mammals, and discuss the significance of the four-chambered heart.
6. Describe and identify the principal organs of the urogenital system of the rat. Demonstrate differences between male and female rats.
7. Identify the main parts of the brain of the rat or sheep, and discuss the basic organization of the mammalian nervous system.

Introduction

The white laboratory rat, an albino form of the Norway rat *Rattus norvegicus,* is often studied as a representative mammal because it exhibits most of the organs and organ systems typical of mammals. Among the distinguishing features of mammals (Class Mammalia) are a body surface covered with hair, an integument (skin) with mammary and other types of glands, skull with two occipital condyles, seven cervical (neck) vertebrae, teeth borne on bony jaws, moveable eyelids and fleshy external ears (pinnae), a four-chambered heart, persistent left aorta, and a muscular diaphragm separating the thoracic and abdominal cavities. Mammals also are warm-blooded or **homeothermic.** The young develop within the uterus of the female, have a placental attachment for nourishment, and are enveloped by special fetal membranes (amnion, chorion, and allantois). Milk to nourish the young after birth is produced by mammary glands. Other common representatives of this group include moles, bats, whales, mice, deer, monkeys, horses, cattle, and humans.

Classification

Phylum Chordata

Subphylum Vertebrata

Class Mammalia

Order Rodentia

Suborder Myomorpha

Family Muridae

Genus *Rattus*

Species *norvegicus*

Materials List

Preserved specimens
  White rat, double or triple injected
  Sheep brain
  Beef heart
Plastic or liquid mounts
  Rat dissection
Prepared microscope slides
  Rat testis, cross section
  Rat kidney, median section
  Rat heart, longitudinal section
Other
  Mounted rat skeleton
  Mounted cat skeleton

External Anatomy

■ Select a preserved specimen and place it in a dissecting pan. Observe the two principal external features that distinguish the rat as a mammal, the **hair** covering most of the body and the paired **mammary glands.** Locate the paired nipples or **teats** on the ventral surface of the trunk between the forelimbs and hindlimbs. How many pairs are present? Are they present on both male and female rats?

The body of the rat is divided into an anterior head connected to the cylindrical trunk by a short, thick neck. A long tail extends posteriorly from the trunk. Interiorly, the trunk is divided by a muscular **diaphragm** into an anterior **thorax** and a posterior **abdomen.** The diaphragm is another distinguishing feature of mammals.

■ Observe the long tail. How does its exterior surface differ from that of the remainder of the body?

Study the cone-shaped head which has an elongated face. On the head, find the two **eyes** and two **ears.** The eyes have upper and lower eyelids and a reduced third eyelid or **nictitating membrane** which can be found on the medial portion of the eye opening beneath the two outer lids.

Note that each ear has an external fold of tissue called the **pinna** which aids in directing sound waves to the opening of the ear, the external auditory meatus. Other important sense organs located on the head are the **vibrissae,** long sensory hairs that provide the rat with a very effective sense of touch. The vibrissae can tell a rat in an instant if a hole is large enough for him to crawl into. Note that most of the vibrissae are attached to the upper lip. Where else do you find vibrissae?

Observe the well-developed upper and lower **lips** surrounding the mouth. In the center of the upper lip, find the **philtrum,** a groove or cleft separating the lip into right and left halves. Locate the two **external nares** above the upper lip. Inside the mouth find the two long, sharp **incisor teeth** that are characteristic of rodents. These two incisors grow continuously, and the rat wears them down by its gnawing habits. Molars or grinding teeth are found farther back in the jaw and will be studied later.

Observe one of the **forelimbs** and note that the limb consists of an upper portion, a lower portion, and a hand-like portion much like the structure of your own arm. Most of the bones in the rat forelimb are homologous with those of a human. Embryological and anatomical studies have shown that these bones have the same origin in the rat and in the human.

Locate the horny **claws** at the tip of each digit and the walking pads in the palm area. These pads are less well developed in the rat than the cat and catlike animals because rats tend to walk on the digits rather than on the entire palm or sole.

Near the base of the tail locate the **anus.** You can determine the sex of your specimen by studying adjacent urogenital structures. Male rats have a **scrotum** ventral to the anus, which holds the two **testes** during reproductive season. At other times, the testes are suspended inside the abdominal cavity and the scrotum is an empty sac. Anterior to the scrotum is the **penis** with the opening of the male urogenital system at the end of the penis. The penis is usually withdrawn into a sheath of skin, the **prepuce.**

If you have a female rat, find the **vaginal opening** ventral to the anus and the separate opening of the **urethra** in front of the **clitoris,** which is located ventral to the vagina.

Skeletal System

The skeleton of all vertebrates exhibits a common basic pattern, and an examination of the skeletons of a frog, a rat, a cat, and a human reveals many fundamental similarities. Since the bones of the rat are small, biology and zoology students often use a cat skeleton to learn more about the organization of the mammalian skeleton.

■ The skeleton of the rat is illustrated in figure 17.1. Study a mounted cat skeleton and locate the major parts of the skeleton with the aid of figure 17.2 and compare it with the illustration of the cat skeleton.

Muscular System (Optional)

A thorough study of the muscular system of the rat would take more time than is available in most introductory zoology courses. In this exercise, we will limit our study to a few of the major muscles of the ventral surface of the thorax and shoulders to provide an example of the organization of the muscular system.

■ Before you start your dissection, you should review the information in Chapter 16 about the structure and action of skeletal muscles and be sure that you understand the meaning of such terms as **origin, insertion, action, antagonistic muscles, extension, flexion, abduction,** and **adduction.**

Skinning

■ Prior to the study of the muscles, you must first skin the appropriate portion of the body. Most preserved and injected rats have an incision on the midventral surface of the neck where they were injected. Start with this opening and free the skin from the underlying muscle and continue the incision posteriorly, stopping just anterior to the anus as shown in figure 17.3. Extend the incision anteriorly to the mouth and then cut around the side of the mouth.

Fig. 17.1 *Rat skeleton.*

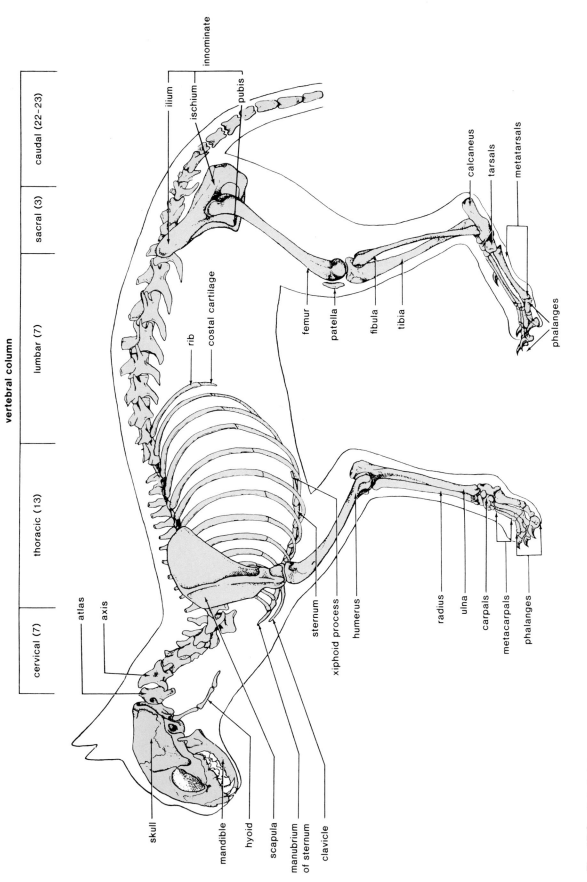

Fig. 17.2 *Cat skeleton.*

■ Next, make two incisions in the chest region between the forelimbs at right angles to the first incision. Extend these lateral incisions to the wrist of each forearm, as indicated in figure 17.3. Carefully separate the skin from the underlying tissue with a blunt probe to expose the superficial (surface) muscles of the trunk and shoulder regions. You may later wish to make similar lateral incisions on the surface of the hindlimbs to study them in more detail.

■ Locate and study the representative muscles listed on table 17.1 with the aid of figure 17.4. Carefully separate each muscle from adjacent ones with a blunt probe and identify its origin (fixed end), insertion (moveable end), and its action (what it does or what it moves).

Internal Anatomy

Mouth and Pharynx

We will begin our study of internal organs with the mouth and pharynx.

■ Cut the muscles, skin, and bones at the corner of the mouth on each side with a pair of scissors or bone cutters. Push down the lower jaw and examine the interior of the **oral cavity.** Locate the muscular **tongue** attached at the rear of the oral cavity.

The anterior part of the tongue is attached ventrally with a thin sheet of tissue, the **frenulum.** Rats have two types of teeth, **incisors** (cutting teeth) and **molars** (grinding teeth). How many of each do you find? Three pairs of **salivary glands** empty into the oral cavity and provide lubrication for the food mass. These glands are the **parotids** (most lateral), the **mandibular glands** (most posterior), and the **sublingual glands** (anterior and lateral to the mandibular glands).

Find the **hard palate** covering the anterior part of the roof of the oral cavity and the **soft palate** covering the posterior portion. The **pharynx** is located at the rear of the oral cavity and consists of three portions: the **oropharynx,** which is below the soft palate; the **nasopharynx,** which receives air from the external nares lies above the soft palate; and the **laryngopharynx,** which is located at the rear and connects with the esophagus.

On the floor of the laryngopharynx find the **glottis,** a slitlike opening into the trachea, which is covered by a small flap of tissue, the **epiglottis.** Opening into the nasopharynx are the **internal nares** and the **eustachian tubes;** the latter connect with the middle ears. The nasopharynx leads to the more posterior laryngopharynx.

Fig. 17.3 *Rat, ventral view, incisions for skinning.*

Other Internal Organs

■ To study other features of internal anatomy, you must cut through the ventral muscles to expose the organs within the coelomic cavity. Make an incision through the muscles just to the right of the linea alba (ventral midline). Start just anterior to the anus and continue anteriorly to the neck region. Cut through the pectoral girdle and pull apart the ribs in the chest region

Table 17.1   Major Muscles of the Ventral Surface and Shoulders

Muscle	Location	Origin	Insertion	Action
Rectus abdominis	Paired muscles adjacent to midventral line of connective tissue (linea alba)	Pubis	First rib, clavicle, and sternum	Supports abdominal organs
External oblique	Flat sheet covering ventral surface of trunk	Ribs and spinal fascia	Ilium, pubis, linea alba	Supports abdominal organs
Internal oblique	Flat sheet lying under external oblique	Spinal fascia, ilium	Cartilages of false ribs, linea alba	Supports abdominal organs
Pectoralis major	Large triangular muscle lying midventral to shoulder girdle	Sternum	Humerus	Adducts forelimb
Pectoralis minor	Posterior and beneath pectoralis major	Sternum	Humerus	Adducts forelimb
Biceps brachii	Anterior ventral surface of humerus	Scapula	Radius	Flexes lower part of forelimb
Triceps brachii	Posterio-dorsal surface of humerus	Humerus	Ulna	Extends lower part of forelimb
Sternomastoideus	Side of neck anterior to pectoralis major	Sternum	Mastoid process of skull	Turns head sideways or back of head down
Sternohyoideus	Medial to sternomastoid	Occipital bone of skull	Hyoid bone	Elevates hyoid

to expose the heart and lungs. Make two transverse cuts through the skin and superficial muscles on each side just behind the pectoral girdle and another pair of cuts just in front of the pelvic girdle (see figure 17.5). Take care not to damage the organs inside the abdominal cavity beneath the superficial muscle layers. Free the diaphragm from the body wall and pin aside the skin and body wall to expose the abdominal organs as in figure 17.5.

Digestive System

The digestive system consists of the mouth, oral cavity, pharynx, esophagus, stomach, small intestine, large intestine, rectum, and terminates at the anus. Several parts of the digestive system are shown in figures 17.5 and 17.6. The **esophagus** is a cylindrical tube leading from the pharynx to the stomach. It enters the **stomach** on the **lesser curvature** (medial concave surface). The **greater curvature** is the larger convex lateral surface. Trace the path of the esophagus (dorsal to the trachea) on your specimen to its entrance into the stomach. The stomach lies beneath the liver on the left side of the abdomen and consists of three portions: a **forestomach** (fundus) that serves mainly for temporary storage; a glandular **middle portion** which secretes mucus, hydrochloric acid, and pepsin (a protease);

and a smaller posterior **pyloric region** with a muscular **pyloric valve** that controls passage of material into the small intestine.

The small **intestine** consists of three specialized regions: the **duodenum,** the **jejunum,** and the **ileum.** The anterior portion of the small intestine is the duodenum. It is 10 to 12 inches in length and receives ducts from the liver and pancreas. Most digestion begins here. Behind the duodenum is the shorter jejunum. The remaining portion of the small intestine posterior to the jejunum is the ileum. The exterior surface of the ileum often feels lumpy because of the presence of numerous lymph nodes embedded in its walls.

Locate the large **liver,** which has four lobes. Bile ducts from each of the lobes join to form a single duct carrying bile to the duodenum, but rats typically lack or have a reduced gallbladder. Anterior to the liver, locate the flat, muscular **diaphragm** (figure 17.6), which partitions the coelom into the thoracic and abdominal cavities. The **pancreas** is not a discrete organ in the rat, but consists of scattered patches of white or pinkish glandular tissue embedded in the mesentery (thin sheet of connective tissue) between the duodenum and the stomach.

sternohyoideus

sternomastoideus

biceps brachii

triceps brachii

latissimus dorsi

external oblique

internal oblique

pectoralis major

pectoralis minor

rectus abdominis

linea alba

Fig. 17.4 *Rat muscles, dorsal view.*

maxillary
gland

trachea

lung

liver

diaphragm

small
intestine

large
intestine

clitoris

anus

Fig. 17.5 *Rat internal organs, female, ventral view. (Photograph by Ken Taylor.)*

The ileum empties into the large intestine, or colon. At the junction of the ileum and colon, find the **caecum,** a blind sac that extends posteriorly from the **ileocolic valve** between the ileum and colon. From this junction, the colon extends forward as the **ascending colon.** The **transverse colon** extends across the abdominal cavity adjacent to the diaphragm, and, in the region of the stomach, it loops back as the **descending colon.**

The caecum and colon are important in resorption of ions and water from the gut contents. Also, numerous mucus glands are found in the epithelium of the colon, which facilitate movement of the contents. The terminal portion of the colon is the muscular **rectum,** which leads to the **anus.**

Respiratory System

During respiration, air enters the external nares and passes through the nasal passages to the nasopharynx and downward through the glottis into the **larynx.** The air is warmed and filtered during its passage through the nasal passages. The presence of vocal cords in the lateral walls of the larynx allows rats to make audible sounds and makes possible some primitive sound communication.

From the larynx, air passes to the **trachea,** a hollow tube supported by a series of incomplete cartilage rings (figure 17.7). Trace the path of the trachea from the larynx to the thoracic cavity where it branches into **two bronchi.** The bronchi enter the lungs, branch further into many **branchioles,** and terminate in many highly vascularized **alveoli** where the gas exchange takes place.

Circulatory System

The circulatory system of the rat exhibits the same basic pattern characteristic of all mammals, including humans, with two major features: (1) a **four-chambered heart,** and (2) two separate divisions for circulation of the blood—a **systemic circuit** and a **pulmonary circuit.**

The systemic division carries blood from the heart through a branching system of arteries to most organs and tissues of the body and collects blood from these organs via a system of capillaries and veins, and brings the blood back to the heart. The pulmonary division carries oxygen-poor blood from the heart to the lungs for oxygenation and returns the oxygenated blood to the heart.

This pattern of circulation shown by mammals represents a major advance over the patterns exhibited by lower vertebrates such as the shark or the bony fish, which have a two-chambered heart and a single circulatory circuit, and the frog, which has a three-chambered heart and an incomplete double circuit which allows some mixing of oxygenated and deoxygenated blood in the heart.

In mammals, deoxygenated blood from the body (except the lungs) returns to the right atrium by the large anterior vena cava and posterior vena cava (figure 17.8). From the right atrium, blood goes to the right ventricle and is pumped via the pulmonary arteries to the lungs to be oxygenated.

Oxygenated blood from the lungs passes through the pulmonary veins to the left atrium of the heart. From the left atrium, the blood goes to the left ventricle and is pumped through the aorta and its branches to all parts of the body except the lungs.

The best specimens for study of the anatomy of the circulatory system are preserved rats that have been injected with colored latex. Specimens may be single, double, or triple injected. Single-injected specimens have red latex injected into the arterial system. In double-injected specimens, the arteries are red and the veins are blue; triple injected specimens are similar to double-injected specimens, and also have yellow latex injected into the hepatic portal system. (As you might expect, the latter are the most expensive.)

Heart and Arterial Circulation

■ Carefully remove the pericardial membranes surrounding the heart. Take care not to break the blood vessels attached to the heart. Study figure 17.8 and locate the two small, saclike **atria** lying on the surface of the two muscular **ventricles.** Note that the ventricles are unequal in size; which ventricle is the larger in your specimen? Examine a microscope slide of heart tissue to observe the difference between the thick, muscular ventricles and the thin-walled atria.

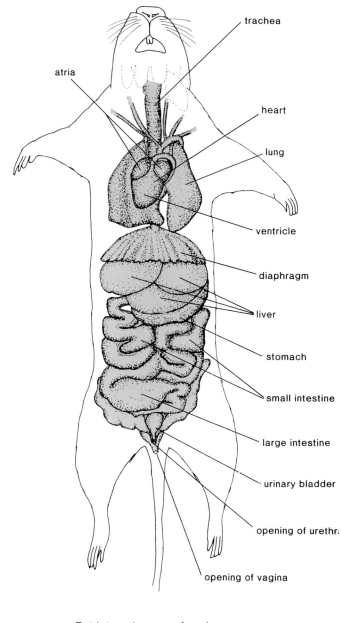

Fig. 17.6 *Rat internal organs, female.*

Extending forward from the heart, locate the large **aorta,** which exits from the left ventricle, curves to the left, and passes dorsal to the heart (figure 17.7). This large curvature forms the **aortic arch,** which gives off five branches. The first two branches of the aortic arch are two **coronary arteries,** which supply the muscles of the heart. They arise within the heart and are often difficult to locate. The other three branches of the aortic arch are more prominent and are easier to locate. These branches are the **innominate artery,** the **left common carotid artery,** and the **left subclavian artery** (figures 17.7, 17.8, and 17.9).

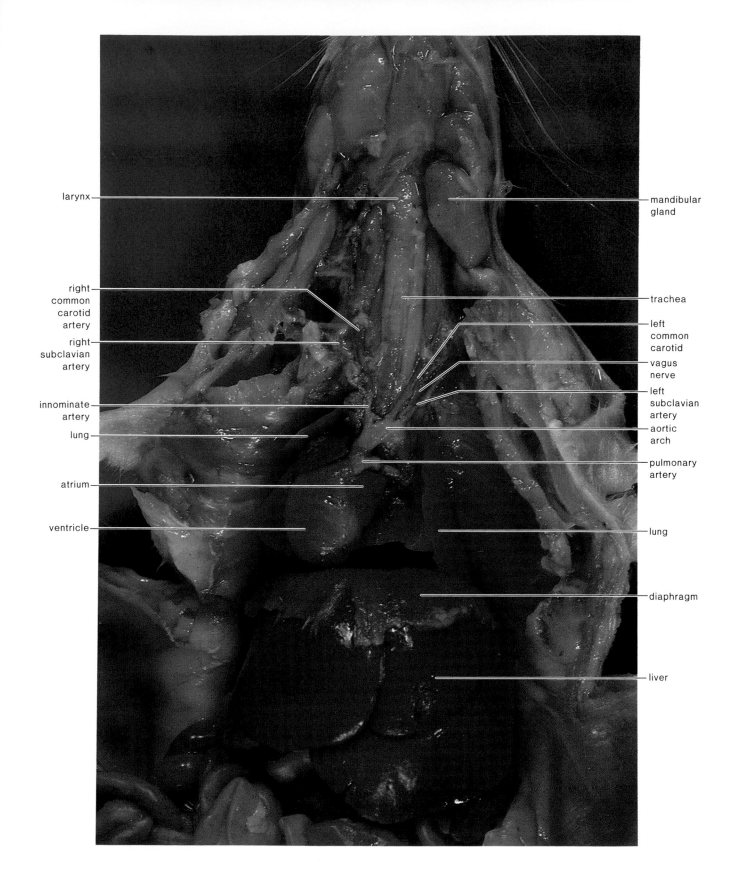

Fig. 17.7 *Rat neck and heart regions, ventral view. (Photograph by Ken Taylor.)*

larynx

mandibular gland

right common carotid artery

trachea

right subclavian artery

left common carotid

vagus nerve

innominate artery

left subclavian artery

lung

aortic arch

atrium

pulmonary artery

ventricle

lung

diaphragm

liver

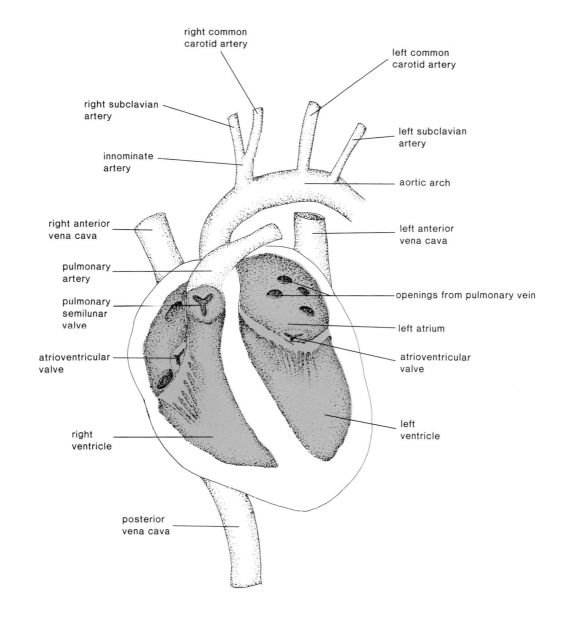

right common
carotid artery

left common
carotid artery

right subclavian
artery

left subclavian
artery

innominate
artery

aortic arch

right anterior
vena cava

left anterior
vena cava

pulmonary
artery

openings from pulmonary vein

pulmonary
semilunar
valve

left atrium

atrioventricular
valve

atrioventricular
valve

right
ventricle

left
ventricle

posterior
vena cava

Fig. 17.8 *Rat heart, longitudinal section.*

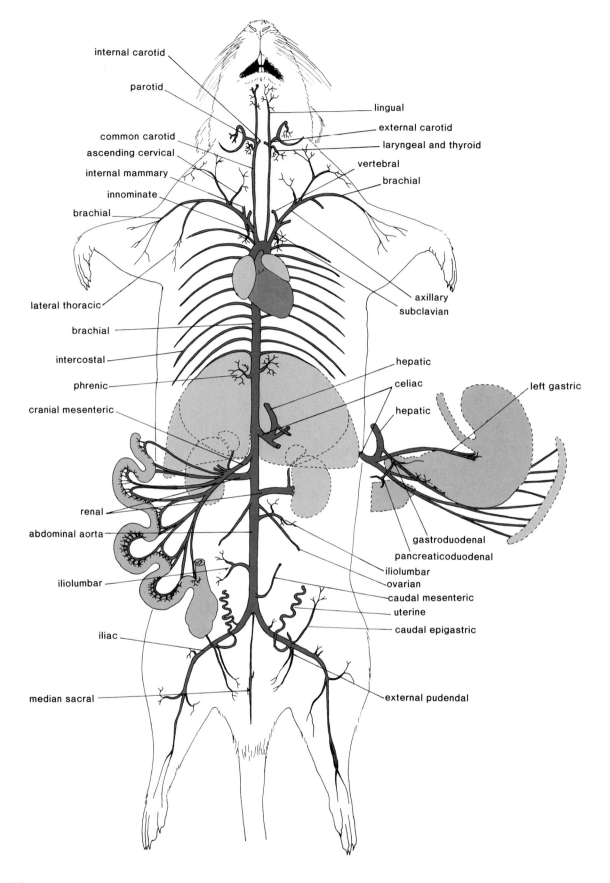

internal carotid

parotid

lingual

external carotid

common carotid

laryngeal and thyroid

ascending cervical

vertebral

internal mammary

brachial

innominate

brachial

lateral thoracic

axillary

subclavian

brachial

intercostal

hepatic

phrenic

celiac

hepatic

left gastric

cranial mesenteric

renal

abdominal aorta

gastroduodenal

pancreaticoduodenal

iliolumbar

ovarian

iliolumbar

caudal mesenteric

uterine

iliac

caudal epigastric

median sacral

external pudendal

Fig. 17.9 *Rat arterial system, ventral view.*

The innominate artery extends forward and divides to form the **right common carotid artery,** which supplies the head and brain, and the **right subclavian artery,** which supplies the right forelimb and shoulder. The **left common carotid artery** branches from the aortic arch to the left of the innominate and passes anteriorly along the left side of the trachea. It gives rise to several branches that supply blood to the left side of the neck and head. The third branch from the aortic arch is the **left subclavian artery,** which supplies the left forearm, the chest, and the neck through several branches.

Posterior to the heart, the aortic arch continues as the **dorsal aorta,** which runs along the dorsal wall of the coelom. Along its course from the arch to the posterior end, the dorsal aorta gives off several major branches. How many of these branches can you locate on your specimen?

Branches of the Aorta Posterior to the Heart

Phrenic arteries—carry blood to the diaphragm
Lumbales arteries—supply back muscles and adrenal glands
Celiac artery—unpaired artery supplying the spleen, liver, stomach, pancreas, and duodenum
Cranial mesenteric artery—unpaired artery to the mesentery and small intestine
Renal arteries—supply the kidneys
Spermatic arteries (male)—supply the testes **or**
Uterine arteries (female)—supply the ovaries and uterus
Iliolumbar arteries—supply dorsal wall muscles
Lumbar arteries—supply back muscles
Caudal mesenteric artery—unpaired artery to colon and rectum
Common iliac arteries—branch several times to carry blood to the hindlimbs, scrotum, and penis (male) or uterus and vagina (female)
Median sacral artery—supplies blood to the tail

Venous Circulation

The pattern of venous circulation consists of four distinct parts: (1) pulmonary circulation from the lungs, (2) collection of blood from the head and anterior parts of the body, (3) collection of blood from the abdomen and posterior part of the body, and (4) the hepatic portal system. We will consider each of these portions of the venous circulation separately.

Veins are thin-walled vessels that carry blood back to the heart after it has traversed capillary beds in the tissues. Veins are filled with blue latex in injected specimens, but because of their thin walls some veins burst from injection pressure, and often smaller veins may be hard to find because they are not well injected.

Pulmonary Circulation. Oxygen-depleted blood leaves the right ventricle of the heart through the large pulmonary arch, which divides into the left and right pulmonary arteries just dorsal to the heart. After the release of carbon dioxide and uptake of oxygen in the capillaries surrounding the alveoli in the lungs, blood is returned to the heart via the right and left pulmonary veins.

Systematic Veins. System blood returns to the heart from the head, neck, forelimbs, and thoracic regions via a pair of large **anterior vena cava veins,** which empty into the right atrium (figure 17.10). The best method for studying the venous system is to start at the right atrium and trace the veins away from the heart. This allows you to start with the larger veins that are usually best injected and therefore are easier to find.

Locate first the left and right anterior vena cavae, which empty into the right atrium. Each of these anterior vena cava veins is formed at the level of the first rib by the junction of three veins: the **internal jugular,** the **external jugular,** and the **subclavian.** The internal jugular receives blood from the trachea, larynx, and neck. The external jugular receives blood from the shoulder and the head. The subclavian receives blood from the forearm and part of the shoulder. Locate these veins on your specimen and observe how they come together to form the anterior vena cava.

The **posterior vena cava** is a single vein extending from the posterior end of the abdominal cavity to the heart. Locate the attachment of this vein to the heart. What other blood vessel enters the heart with the posterior vena cava?

Several other vessels join the posterior vena cava in the abdominal cavity. Locate the following:

Hepatic veins—collect blood from the liver. How many do you find in your specimen?
Phrenic veins—enter the vena cava where it passes through the liver.
Renal veins—collect blood from the kidneys (also from adrenal glands and left gonad).
Right gonadal vein—collects blood from the right gonad.
Iliolumbar veins—carry blood from the dorsal lumbar region.
Common iliac veins—large paired veins that join to form the posterior vena cava.

Each common iliac vein collects blood from several smaller veins from the hindlimbs and posterior abdomen, including the pelvic veins from the pelvic region, the caudal veins from the tail, and the femoral veins from the hind limbs.

Hepatic Portal System. A special venous system in the rat collects blood from parts of the digestive tract and carries this nutrient-rich blood to the liver. This is the hepatic portal system. (See chapter 16 for a discussion of portal systems.) In the liver, most of these nutrients are removed for storage or chemically changed before they are released into general circulation. The principal vessel is the hepatic portal vein, which carries blood from the intestines to the liver. Find the hepatic portal vein in the mesenteries posterior to the liver. The hepatic portal collects

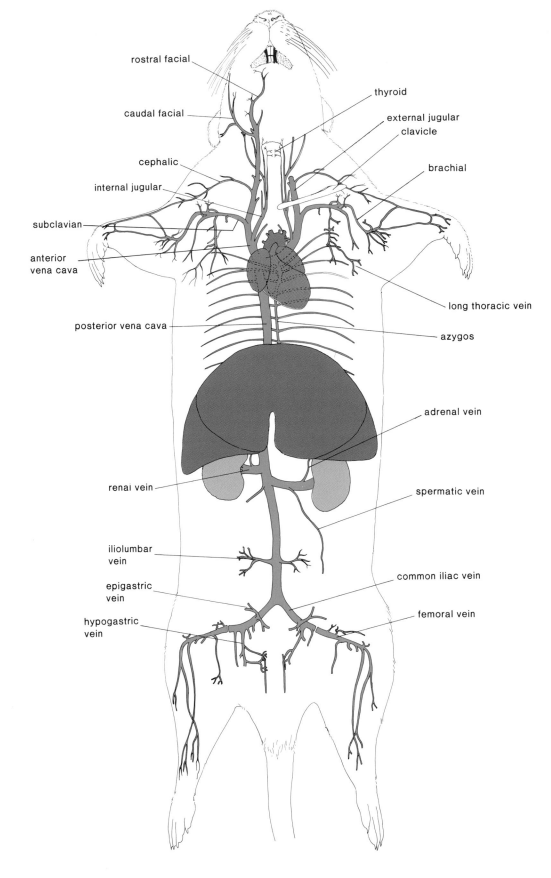

rostral facial

thyroid

caudal facial

external jugular

clavicle

cephalic

brachial

internal jugular

subclavian

anterior
vena cava

long thoracic vein

posterior vena cava

azygos

adrenal vein

renal vein

spermatic vein

iliolumbar
vein

common iliac vein

epigastric
vein

femoral vein

hypogastric
vein

Fig. 17.10 *Rat venous system, ventral view.*

blood from several smaller veins leading from the small intestine, large intestine, stomach, pancreas, and spleen. From the liver, blood passes through the hepatic vein to the posterior vena cava to the heart, as already seen.

Spleen and Thymus

Two accessory organs that have important roles in the circulatory system are the **spleen** and **thymus.** The spleen is a dark, elongated structure attached to the mesentery along the greater curvature (larger convex surface) of the stomach. The spleen serves as a reservoir for the storage of red blood cells and is important in the immune system. It functions also as part of the lymphatic system in filtering out and neutralizing infectious agents.

The thymus in adult rats consists of a small mass of tissue on the ventral surface of the trachea anterior to the heart and near the branching of the two bronchi from the trachea.

The thymus is most well developed in young rats where it serves to produce white blood cells. It decreases in size in older rats. It is made up largely of lymphatic tissue and functions in immunity and other forms of defense against infectious agents.

Urogenital System

The excretory and reproductive organs of the rat are closely related both embryologically and anatomically and are commonly considered together as components of a single system, the **urogenital system.** The chief components of this system involved with excretion are the **kidneys,** the **ureters,** the **urinary bladder,** and the **urethra.** The paired kidneys are closely held to the dorsal wall by a layer of peritoneum that separates them from the abdominal cavity. Attached to the surface of each kidney is a thin sheet of reddish tissue, the adrenal gland. The adrenal glands are endocrine organs that produce several important hormones. The kidneys serve to concentrate nitrogenous wastes of metabolism and produce urine. Urea is actually produced in the liver and is transported to the kidneys. The concentration of wastes and the production of a concentrated urine is an important water conservation adaptation of rats and many other terrestrial animals.

■ Cut through the peritoneum covering one kidney, remove the fat deposits around the kidney, and make a longitudinal section of the kidney with a razor blade. Observe that the kidney tissue consists of an outer **cortex** and an inner **medulla** area, which is darker in color.

The inner concave surface of the kidney where the ureter and blood vessels attach is the **hilus.** Identify the **renal artery** and **renal vein** adjacent to the large **ureter,** which carries the urine to the **urinary bladder.** From the urinary bladder, trace the **urethra** to its external opening adjacent to the clitoris if your specimen is a female. If you have a male specimen, trace the urethra to the penis.

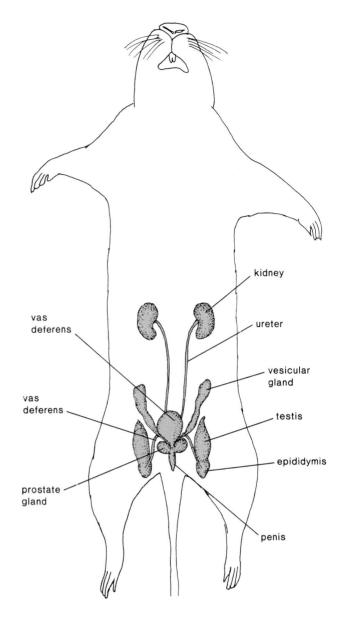

Fig. 17.11 *Rat urogenital system, male.*

Male Reproductive Organs

The male reproductive system is illustrated in figures 17.11 and 17.12. Sperm are produced in the paired **testes.** During breeding season, the testes are located in the **scrotum,** a large sac located ventral to the anus. At other times, the testes are usually retracted through the inguinal canal into the posterior part of the abdominal cavity. Locate the testes in the scrotum or abdominal cavity, and carefully cut through the connective tissue surrounding one testis. Find the **epididymis,** a coiled tube attached to the surface of the testis where mature sperm are stored (figure 17.13). A smaller tubule, the **vas deferens,** carries the sperm from the epididymis to the **urethra.** Follow the vas deferens from one testis and observe that it joins the vas deferens from the other testis where they empty together into the urethra. The urethra carries the sperm from the two vasa

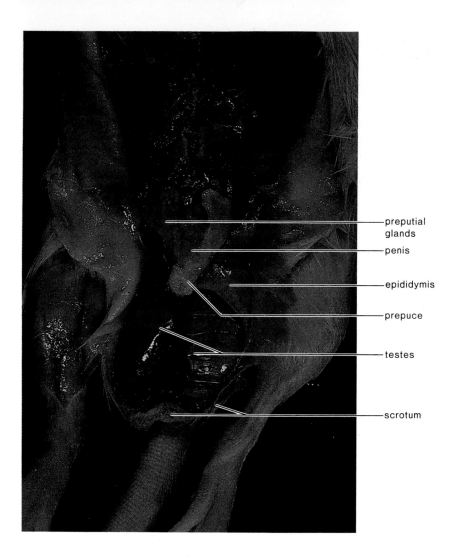

preputial glands

penis

epididymis

prepuce

testes

scrotum

Fig. 17.12 *Rat urogenital system, male, scrotum dissected to show testes. (Photograph by Ken Taylor.)*

deferentia to the penis, from which it is deposited in the vagina of the female during copulation. The sperm are suspended in the seminal fluid or **semen** made up of the secretions of several glands associated with the urethra and penis.

Female Reproductive Organs

The female reproductive system is illustrated in figures 17.6 and 17.14. Eggs or **ova** are produced in the paired **ovaries** found just posterior to the kidneys. The ovary is a small mass of follicles embedded in mesentery rather than in a compact organ. The **oviducts** in the rat are small, short tubules that lead from the funnel-shaped ostia where ova are received from the ovaries. The oviducts connect with the two branches of the **uterus.** The uterus consists of two large branches or horns of the uterus which open separately into the **vagina.** These **uterine horns** represent a significant morphological adaptation that facilitates the simultaneous development of several embryos and mul-

tiple births. The vagina leads from the two uterine horns to the exterior and receives sperm during copulation. Mature ova descending through the oviduct are fertilized in the uteri. The fertilized eggs are subsequently implanted into the uterine wall. The gestation period (time of uterine development before birth) of the laboratory rat is about 21–22 days. A litter averages about 6–8 young.

Nervous System

The nervous system of mammals consists of three major divisions: (1) the central nervous system, consisting of the brain and spinal cord; (2) the peripheral nervous system, made up of the myelinated sensory and motor nerves extending from the central nervous system; and (3) the autonomic nervous system, a system of visceral nonmyelinated nerves that regulate the glands and visceral organs. In this exercise, we will confine our study mainly to the brain and the central nervous system.

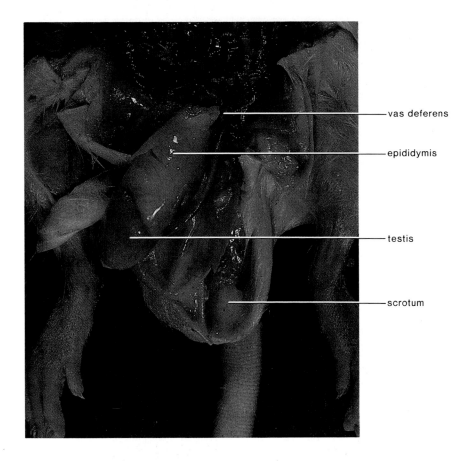

Fig. 17.13 *Rat urogenital system, male, testes dissected to show internal details. (Photograph by Ken Taylor.)*

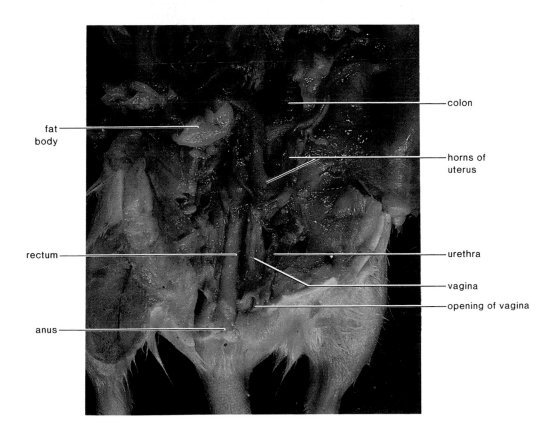

Fig. 17.14 *Rat urogenital system, female. (Photograph by Ken Taylor.)*

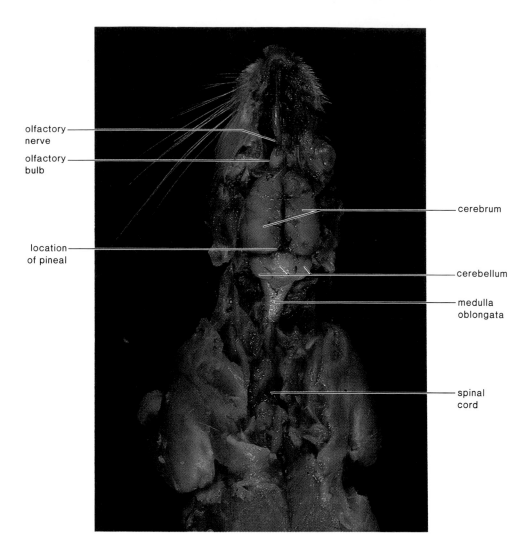

olfactory nerve

olfactory bulb

location of pineal

cerebrum

cerebellum

medulla oblongata

spinal cord

Fig. 17.15 *Rat brain, dorsal view. (Photograph by Ken Taylor.)*

Table 17.2   Principal Regions of the Mammalian Brain

Region	Principal Structures
Telencephalon	Olfactory bulbs, cerebral hemispheres
Diencephalon	Pituitary gland
Mesencephalon	Corpora quadrigemina
Metencephalon	Cerebellum, pons, part of medulla oblongata
Myelencephalon	Part of medulla oblongata

Brain

The brain of the rat is small and is more difficult to study than that of larger mammals. We will therefore supplement our study of the rat brain with the larger but similar brain of the sheep.

■ Remove the skin and hair from the top of the head between the eyes and ears, and cut away the muscles at the back of the head near the occipital region. Use bone shears to cut away a portion of the bone at the back of the skull to expose the brain. Take care not to cut into the brain. When you have exposed the brain, use forceps to pick away small pieces of bone to further expose the brain (figure 17.15).

Observe that the brain is covered by several tough outer layers of connective tissue, the meninges. The brain of the rat and other mammals consists of five principal regions that can be traced to its embryological development. These anatomical regions and the principal structures in the adult brain representing these divisions are listed in table 17.2.

dorsal

ventral

Fig. 17.16   *Rat brain, dorsal and ventral views.*

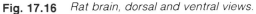

Locate first the two large anterior **cerebral hemispheres** separated by a median groove or fissure. Anterior to the cerebral hemispheres find the two smaller **olfactory bulbs** to which the olfactory nerves connect (figure 17.16). At the posterior margin of the cerebral hemispheres in the cerebral fissure is the **pineal body.** Posterior to the cerebral hemispheres on the dorsal surface find the **cerebellum,** consisting of three lobes, right and left hemispheres, and the vermis, each with a folded surface. The cerebellum is the principal center responsible for coordination of body movements.

The **medulla oblongata** is the most posterior part of the brain and connects with the spinal cord. The medulla regulates such vital processes as respiration, heart rate, blood pressure, and hormonal secretion.

▣ After you have studied these structures on the dorsal surface of the brain, carefully detach the brain from the braincase. Try to retain a portion of the several

pairs of cranial nerves attached to the ventral surface of the brain so that you can locate where these nerves arise from the brain.

On the ventral surface of the brain, locate the anterior **olfactory bulbs** with two large olfactory tracts leading posteriorly to the cerebrum and the **optic chiasma** where the two large **optic nerves** cross. Posterior to the optic chiasma is the **pituitary gland,** which may be left on the floor of the brain case when the brain is lifted out. Behind the pituitary, a transverse tract of nerve fibers, the **pons,** connects the left and right hemispheres of the cerebellum.

Twelve pairs of **cranial nerves** arise from the ventral surface of the brain (figure 17.16). We have already noted olfactory tracts and the optic nerves, which are the first and second pair of cranial nerves. The twelve pairs of cranial nerves and their locations are listed in table 17.3.

Table 17.3 Cranial Nerves of the Rat

No.	Name	Location	Function(s)
I.	Olfactory	Olfactory bulbs	Smell
II.	Optic	Diencephalon	Vision
III.	Oculomotor	Ventral surface of mesencephalon	Eye movements
IV.	Trochlear	Dorsal surface of mesencephalon	Eye movements
V.	Trigeminal	Pons	Innervates skin, vibrissae, jaw muscles, tongue, and teeth
VI.	Abducens	Medulla oblongata	Eye movements
VII.	Facial	Medulla oblongata	Innervates jaw muscles
VIII.	Auditory	Medulla oblongata	Hearing
IX.	Glossopharyngeal	Medulla oblongata	Innervates pharynx and tongue
X.	Vagus	Medulla oblongata	Innervates larynx, heart, lungs, diaphragm, and stomach
XI.	Accessory	Medulla oblongata	Innervates neck muscles and pharyngeal organs
XII.	Hypoglossal	Medulla oblongata	Tongue movements

Sheep Brain

For further information on the anatomy of the mammalian brain, study a preserved sheep brain (see figure 17.17). Identify the principal structures of the sheep brain and compare them with the rat brain. What differences do you find?

Spinal Cord

The spinal cord extends posteriorly from the medulla oblongata and is covered by meninges. The spinal cord passes through the vertebra in the vertebral column, and several pairs of **spinal nerves** extend from the spinal cord through openings between the vertebrae. The spinal nerves are usually classified in groups according to the region where they arise from the spinal cord: **cervical, thoracic, lumbar, sacral,** and **caudal.**

Demonstrations

> Liquid mount of dissected rat
> Microscope slide with longitudinal section of a rat heart
> Microscope slide with cross section of rat testis
> Microscope slide with cross section of rat ovary
> Microscope slide with median section of rat kidney
> Microscope slide with rat sperm

Key Terms

Appendicular skeleton   portion of the skeleton of the rat and other vertebrates that includes the bases of the forelimbs and hindlimbs, and the bones that provide their articulation (girdles) with the axial skeleton.

Axial skeleton   portion of the skeleton of the rat and other vertebrates that provides support for the main axis of the body; includes the skull and the vertebral column.

Diaphragm   muscular sheet located anterior to the liver that separates the thoracic and abdominal cavities in mammals.

Hepatic portal system   portion of the venous circulation in many vertebrates that collects blood from capillary beds in the stomach and intestine, and returns it to the

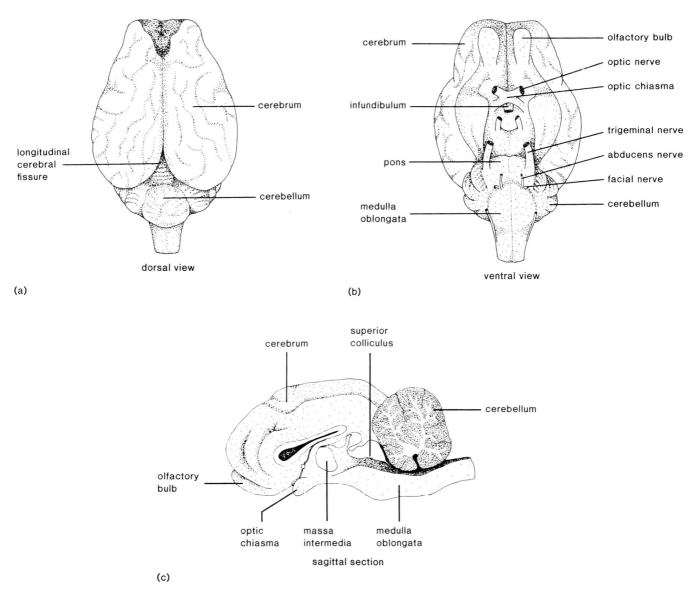

Fig. 17.17 *Sheep brain, dorsal view, ventral view, and sagittal section.*

liver, where the blood passes through a second capillary bed where nutrients are removed prior to returning the blood to the heart.

Mesentery   double sheet of mesodermal tissue that supports various internal organs in the coelom of vertebrates.

Peritoneum   thin sheet of mesodermal epithelium which lines the coelom of vertebrates.

Pulmonary circulation   portion of mammalian circulatory system that carries blood from the heart to the lungs and returns it to the heart.

Systemic circulation   portion of the mammalian circulatory system that carries blood from the heart to and from all organs of the body except the lungs.

Vibrissae   long sensory hairs of the face of the rat.